A SHA

A NOTE ON THE AUTHOR

Joe Shute is an author and journalist with a passion for the natural world. He studied history at Leeds University, and currently works as a senior staff feature writer at The Telegraph. Before joining the newspaper, Joe was the crime correspondent for The Yorkshire Post. He lives with his wife in Sheffield, on the edge of the Peak District.

@JoeShute

A SHADOW ABOVE

THE FALL AND RISE
OF THE RAVEN

Joe Shute

BLOOMSBURY WILDLIFE
LONDON · OXFORD · NEW YORK · NEW DELHI · SYDNEY

 ... WILDLIFE
... g Plc
... C1B 3DP, UK

... nd the Diana logo are trademarks
... ng Plc

... tain 2018
... 2019

Copyright © Joe Shute, 2018
Illustrations © Liz Collins, 2018

Joe Shute has asserted his right under the Copyright, Designs and Patents Act,
1988, to be identified as Author of this work

A catalogue record for this book is available from the British Library.

Library of Congress Cataloguing-in-Publication data has been applied for.

ISBN: HB: 978-1-4729-4028-5; PB: 978-1-4729-4029-2; ePub: 978-1-4729-4030-8

2 4 6 8 10 9 7 5 3 1

Cover by jasmineillustration.com

Typeset in Bembo Std by Deanta Global Publishing Services, Chennai, India
Printed and bound in Great Britain by CPI Group (UK) Ltd, Croydon CR0 4YY

To find out more about our authors and books visit www.bloomsbury.com.
and sign up for our newsletters

For my wife and family, to whom I owe everything

Contents

Prologue

If you place your hand on the top of your shoulder and press your fingers into the muscle where the clavicle meets the scapula, you will find it: the coracoid process, a small curved piece of bone that holds the joint together and takes its name from the ancient Greek for the raven beak.

Or cast your eyes up to the night sky. The Corvus constellation sits just to the right of Virgo. The brightest of its stars, Gienah, which forms the shape of the raven's wing, is 165 light years from earth.

It is after work, and I am on my way home through the Peak District after three long days away, but the evening is still too early for the stars to show through the faltering light. I am parked by the side of the road, watching a raven that I had noticed foraging among the cotton grass on the roadside verge as I approached. I am not far from my house and know this bird: a haggard, fully-grown adult that I often spot on my way across the moors and like to think has lived here long before me.

As soon as my foot depresses the brake pedal, the raven takes flight with a languid ease, wings creaking into action as if attached to invisible pulleys. As it flies away from me, it pursues what has become a now familiar flight line, skirting the lower branches of a row of old oaks and parallel to a drystone wall bordering a small wooded enclosure. This particular raven tends to prefer to fly close to the capstones, the tips of its primaries almost brushing the curved tops as delicately as a quill dipped in an inkpot.

On the car radio they are talking about Manchester, from where I have just driven over Snake Pass, and the man who walked into a concert venue and detonated a shrapnel bomb in his rucksack, killing 22 people, mostly women and children, injuring 250 others and tearing himself in half.

My journalist colleagues and I had been tasked with piecing together and reporting on the horrors of his crime. I had driven between the bombsite and the vigils and homes of grieving friends and relatives. I had written about flowers and tributes and pink balloons released in memory of victims as young as eight. I had been called a vulture in one pub and been bought a drink in another. As I've been driving home, those images have flickered in my thoughts like a showreel.

The raven comes to rest on a gatepost and sits looking back at me. I rub my tired right shoulder, massaging my fingers deep into the joint around the raven bone where my rucksack had chafed during the days on the road. The last rays of the sun pick out the shimmering, oily, midnight-blue in its plumage. I turn off the radio and wind down the driver's window, hoping to hear the bird call out in the twilight, but it maintains a silent solemnity.

I know this raven's territory, but not where it roosts, nor even whether it has a mate. After we consider each other for a while longer its long, steel-ringed talons skitter off the stones and its wings judder upwards taking it over the trees and out of sight. I contemplate the empty silence of the moor then start the engine to drive back home.

I feel as I always do when I have been watching a raven, a curious sense of realignment. I see in this bird of blood an emblem of my own age: a symbol of its darkness and yet still somehow one of hope; of the rise and fall of empires and the continuum of life; of the wildness we have lost and that which remains within us.

I was born in 1984, making me the flag-bearer of a strange generation. Raised in a comfortable home to a loving family, secure in the knowledge that, as the songs politicians played during the election campaign told us: things can only get

better. And then, just as I had left university and started my
first proper job in journalism on a local paper, along came the
financial crash of 2007; and with it the collapse of all the
misplaced entitlement of my youth. Since then, everything
has changed. Things were getting more dangerous and
unstable; we would never again have it so good. Rather than
better, it was going to get far worse.

I could not say exactly when I started noticing the birds
around me, but it was an interest that sharpened during this
tumultuous time. As the certainties of one world started to
dissolve, I began to delve into another that I had previously
known little about. I suppose it was a release, at first, but I
soon discovered a soothing surety in these ancient rhythms
of migration and breeding; of fledge and moult. Learning
more about birds helped me to become less fearful of my
own world, even as it became an increasingly savage place
to exist in. To be precise, I stopped seeing it simply as
'my world'.

Ravens have always stood out for me. The sheer bulk of a
bird that weighs in at 1.3kg (2.8lb) and possesses a wingspan
of 1.5m (5ft) means you can't fail to be impressed. But aside
from this statuesque presence that manages to encapsulate
both eagle and vulture, there is an inner life to the raven that
fascinates me.

For as long as humans have been on this earth we have
attempted to explain ourselves through this soulful bird,
drawn our maps upon it – celestial and otherwise – sought
meaning and developed a specific and enduring culture
through its twisted shapes and appetites. The raven has been
our companion through the ages, and in Britain at the heart
of its history. The bird with which we share flesh and bone is
a bellwether for the fortunes of our nation; a harbinger of
change and blurry black punctuation mark denoting the fall
and rise of the epochs that have shaped us.

Ravens are coming back to live among us, returning to both the countryside and human settlements after centuries of exile. The more I watch and learn about these birds the more I want to discover the places and stories associated with them. I want to climb the ancient crags where they have nested for centuries, hear the muscle memory of their wingbeats overhead and fill my ears with their mysterious conversations. I want to visit the raven in the furthest extremities of these islands and also where they had once been feared, excised for eternity, and are now hiding in plain sight. I want to join the dots, draw my own maps and understand where and why they carve up territories for themselves and feel the bird driving back towards me. I find a deep thrill in the thought – and sight – of ravens returning. I seek through them a profound, feral meaning missing in modern life.

We have long attributed to ravens the ability to see further into the future than ourselves. In Roman literature and the stories of Ancient Britons the raven was often depicted as a prophet. A few months after the Manchester bombing, in the summer of 2017, a report was published by a group of Swedish researchers who had been studying five captive ravens at Lund University. Through presenting them with various challenges and rewards and monitoring the response of the individual birds to how they hoarded and bartered with food, the research team managed to prove ravens are indeed capable of thinking about the future. For the first time this study confirmed in the raven the power of foresight, an ability previously only documented by scientists in great apes and humans. Now this bird of augury is back among us, I wonder what it sees for our own dark times.

CHAPTER ONE

Coming in from the Cliffs

On Dungeness beach, a storm rolls in. The shingle boils with each washing-machine wave. Late afternoon turns to twilight, and blacker still.

The air hums: with salt spray whipped up by the rushing ocean, and with the noise of man's installation; a platoon of pylons snaked together with cable marching towards the nuclear power station, Dungeness A. Such imposing structures would dominate most other landscapes, but not this vast Kent headland. Dungeness is known as England's only desert, perhaps because it possesses those two great characteristics of the world's most barren stretches: ever-shifting and impossible to sculpt.

Everything, man-made at least, feels on a short lease. Even the nuclear power station, built in 1983, is to throb its last. In 2015 Dungeness A was decommissioned. In perhaps a decade, the hulking grey casement of its sibling Dungeness B will also join the abandoned wooden cottages deposited all along this beach, mere follies of temporary settlement. True, the old lighthouse still gleams, but on foul afternoons like this, its revolving light is easily blotted out by cloud.

Red flags flutter to show that the Ministry of Defence firing ranges flanking the beach are in use. The shingle is marked by rolls of rusting barbed wire snarled up with orange

Sainsbury's shopping bags, old rope and other detritus of wind and sea. Official signs warn me that if I dare to stroll the wrong way they could be the last steps that I ever make.

A family car drives slowly up the dirt track towards the beach. All four doors open – gingerly, at first, before being yanked full on to their hinges by the wind. The parents, two teenage children and a pitch-black Jack Russell spill out shrieking (and barking) and scatter up the shingle bank to look at the churning waters of the English Channel; today coloured a filthy brown. A few moments later they are back in their car seats, with heads buried into bright anoraks and the heating turned on. I watch my own footprints in the sea-slicked stones. They disappear the moment I lift each boot.

The wind roars in my ears. I watch herring and black-backed gulls strafe close to the water, wheeling in and out of the Ministry of Defence exclusion zone as they search for silver fish among the white caps. Then I turn back towards the sheep pasture where a pair of marsh harriers scythe over the water reeds that fringe an official bird sanctuary from the scrubland. I scan the pylons for another resident of the wilds of Dungeness, a bird previously driven from this landscape. As with so much here, human intervention has only temporary effect.

Ravens are back in Dungeness, restored to the south coast and beyond. The grand bird of myth and mystery has returned after centuries of persecution. And not just to the country's furthest outposts. There has been a 45 per cent increase in the number of ravens in Britain since 1995 and a 121 per cent increase in England over the same period. There are now well in excess of 12,000 breeding pairs across the country, and the black wings of the raven beat ever closer towards our towns and cities.

This is how we used to live; in the company of these great birds in life and in death. Ravens feature prominently in the legends of the ancient Celts, linked to glory, deities and the afterlife. Archaeological surveys of several Celt and Roman sites in Britain have discovered raven bones were more numerous than any bird apart from domestic fowl. Common raven and human remains have also been found commingled in ancient settlements (4,000–10,500 years ago) at Troy in Mesopotamia and modern day Syria, Poland and Western Canada. The Viking settlers who invaded these shores from distant lands across the North Sea sported ravens on their shields and attributed to the birds supernatural powers of augury in their stories of gods and war. In 1066, a few miles from Dungeness, William the Conqueror and his men waded ashore wielding raven banners that had been blessed by the Pope (they even feature on the Bayeux Tapestry).

Birds of omen, but birds of purpose too. During the medieval period, ravens would throng the streets of Britain picking carcasses clean – not discriminating between man and beast. As with our invaders, we came to revere them. In the winter of 1496, an ambassador from Venice who was staying in London wrote of his British guests: 'Nor do they dislike what we so abominate, crows, rooks and jackdaws; and the raven may croak at his pleasure, for no one cares for the omen; there is even a penalty attached to destroying them, as they say that ravens will keep the streets of the town free from all filth.'

By the middle of the fifteenth century, ravens and red kites were officially recognised for the service they provided scavenging rotten meat. It was made a capital offence to kill them. The act that protected them was the first piece of nature conservation legislation designed for the public good, rather than purely to protect hunting rights. In 1534 King Henry VIII issued a special decree protecting ravens hunted for sport by falconers.

Yet this state of harmony could only last so long. By the late seventeenth century the relationship between man and raven had soured. The sight of the birds scavenging the corpses left in the aftermath of the Great Plague of 1665 was supposedly one of the moments when we decided we could no longer live in the company of ravens. Perhaps not helped by the fact the hooked, plague-doctor masks, designed to filter out the putrid air, so closely resembled the raven's beak.

Corvids in general and ravens in particular (the largest of the family), came to be seen as vermin and were driven out with such a vengeful hatred that bounties were placed on the heads of individual birds. Take the testimony of the gentle chronicler of the Hampshire countryside, the Reverend Gilbert White. In his collection of letters, *The Natural History of Selborne*, published in 1789, White details the fate of the birds that occupied an ancient oak in the parish, known as 'the raven tree'.

> The saw was applied to the butt, the wedges were inserted into the opening, the wood echoed to the heavy blows of the beetle or mallet, the tree nodded to its fall; but still the raven dam sat on. At last, when it gave way, the bird was flung from her nest; and though her parental affection deserved a better fate, was whipped down by the twigs which brought her dead to the ground.

From the New Forest to the West Riding of Yorkshire, parish after parish across Britain holds records of the last of its ravens being wiped out. By the twentieth century, raven populations had been decimated to the point where what some called a 'cultural gap' had developed across the country. In many places they had simply vanished.

★ ★ ★

My own interest in birds began as a student in my early 20s with long walks over the West Yorkshire Moors, where my footsteps sent, what I now know to be skylarks and, what even my untrained eye back then realised were grouse, bursting up around me. The lark song was a sound of simple joy: a beat not dissimilar to the drum and bass music we danced to at night.

During trips heading further north into the Yorkshire Dales, I began to understand the pleasure of correctly identifying a lapwing's sculpted plumage and a curlew's bent beak. Driving back to Leeds, past the stately manor Harewood House, red kites soared on the torrents above waiting to pounce on the corpse of an unfortunate rabbit or fox that met their death on the road. Sometimes a bird would swoop right down in front of me, risking the brunt of my turquoise Nissan Micra but allowing me the chance to stare into its orange gimlet of an eye, which possessed an instinct and history that I could not possibly understand. When I started to take special detours past the Greggs bakery in Leeds to watch the fat starlings gobble up pastry crumbs dropped on the street outside, I knew my interest was turning into something else.

I came to birds late. I came to ravens later. I had only known of them through school trips to the Tower of London, where I had watched the fat, glorious specimens wobbling around unsteadily on their claws. I knew the legend; that were the last of these captive ravens to die, then England would fall. I knew too, from the foul glance my teacher shot me as I proffered one of them a crisp from my lunch box, that we were NOT to feed the birds.

Many more years would pass before I saw my first one in the wild. It was in Upper Wensleydale on a winter's day, its black body framed against a white sky. The bird was flying high, and lacking a pair of binoculars, at first, I barely afforded it a second glance, presuming it to be a carrion crow heading home to roost along the valley. But something about the manner of its flight kept me looking. Its wings flapped in an ungainly fashion and with curious jerks – stately and yet somehow otherworldly. The movement reminded me of an old black-and-white Japanese horror film I had once seen where the ghost moved at crackly shutter-speed towards its victims. One moment there, one moment not. The raven's call, when it came, confirmed this was a very different beast: a guttural sound beyond a croak, like wood being scraped along a rock. I was entranced.

Ever since I have sought out ravens wherever I can. A few winters ago I climbed Ben Nevis to find a pair waiting in the snow 1,343m (44,000ft) up. My yelp of excitement that prompted them to lazily take off and vanish into the mist remains a source of regret today. These sightings have often been fleeting and yet have stayed with me for long afterwards. Vast and yet shy – of humans at least, ravens seem far less troubled when I have watched them being dive-bombed by peregrine falcons, merely tipping a wing to allow the fastest animal in the world to shoot past.

These epic aerial battles are now playing out in Dungeness too. Both ravens and peregrines have started nesting on the top of the old substation, which squats like a great steel toad on the shingle banks. The site is guarded around the clock by armed police, so both families of birds only have each other to fear rather than the thieves who still target the eggs of these great wild birds.

If you saw a raven on Dungeness during the mid-1990s you had to write an official report to the County Rarities Committee to have the sighting confirmed. The first raven scouts were spotted flapping over the marshland in 2006, and five years later came the first confirmed breeding on the substation. Ravens have reared their young here every year since, and are among some three to five breeding pairs between Hastings and Dover. At Dungeness, the birds have once more become part of this alien landscape.

But not the day I come looking. In vain I scour the nuclear power station – as far as the authorities allow strange men with binoculars to stray – and wander over the marshland until the mud sucks at my boots. When the rain begins, a kindly birdwatcher in a jeep takes pity on me and allows me to sit in the passenger seat while we discuss ravens and their capacity to infuriate. As we talk the rain only sets in further, drumming on the roof over our heads. I bid him farewell and begin the long trudge back to my car. That the binoculars are still hanging around my neck is more of an afterthought than an indication of any real hope.

And then through the gloom, I see that now familiar black shape. Its wings are hunched like a boxer shrugging off blows, and its mammoth, rhino-horn beak and straggly beard are set dead straight in the direction of home. It is gone in less than a moment. My raven flaps through one of the pylons and then banks right and disappears out of sight behind the shingle bank.

What depth of memory lies in those obsidian eyes? *Corvus Corax*, also known as the common raven, is our largest passerine bird, one of 11 subspecies and the most widely distributed of all corvids, stretching from the Arctic to the Mediterranean and far beyond. Adult birds are nearly twice the size of their close cousin, the carrion crow.

Ravens have lived on our islands since the last Ice Age. As glaciers melted into tundra 10,000 years ago the raven spread northwards. The raven evolved from these boreal wildernesses a fierce and opportunistic omnivore: feathers scaled like plates of armour; its beak several inches long from the bristled base to the hooked tip; possessing the sharp curvature of a bowie knife.

In flight though, all that bulk dissipates. The raven's powerful flight muscles (pectoralis and supracoracoideus)

perfectly balance its keeled sternum pulling its wings up and
down with a grace that defies even the fiercest winds. The
bird is a true aeronaut; it can twist, dive and accelerate with
all the force of a swooping falcon, and soar hundreds of feet
above, coasting upon thermals with its saw-blade flight
feathers. Despite this natural poise, sometimes over fields and
human settlements, it flies as ragged as a harpy. An emissary
of the wild and shadow on the land.

* * *

I return home that night to an email from a man called Paul
Holt telling me that very same afternoon a pair of ravens had
been displaying over the White Cliffs of Dover – just a few
miles from where I had been.

Situated between Folkestone and Dover and built out of the 4.9 million tonnes of chalk marl scooped out of the ground to create the Channel Tunnel, even to get to Samphire Hoe one must drive down a steep tunnelled road flanked by razor wire on the coastal side guarding the entrance to the route to France. As with Dungeness, the Hoe must be on one of England's most heavily guarded nature reserves. The site opened in 1997, three years after the official unveiling of the tunnel, and Paul has been here since the start.

The spoil that created Samphire Hoe means it juts out over the water, a floating man-made island overlooked by the mammoth Shakespeare Cliff, whose sheer chalk face is peppered by an abundance of rock samphire. The cliff takes its name from the famous playwright who visited this area numerous times and used it for a scene in *King Lear* where the blinded Earl of Gloucester is hurled off the top.

Later in the play, Shakespeare once more describes the cliff and 'the crows and choughs that wing the midway air'. He makes no mention of ravens roosting in its cracks and crevices, yet throughout his body of work the birds appear some 50 times, more than any other animal, often as omens of evil portent. In *Hamlet*, for example, 'the croaking raven doth bellow for revenge'. *Othello*, in conversation with Iago, describes a memory returning to him 'As doth the raven o'er the infected house, Boding to all …'. The playwright's ravens act as vehicles of change, of human lives toppled by forces beyond their control.

On Shakespeare Cliff in the Victorian era, heavy industry supplanted the samphire pickers and put paid to the nesting idyll of the ravens. In January 1843, what was then the largest man-made explosion in the world was set off on Shakespeare Cliff by the South Eastern Railway Company to extend the construction of its flagship tunnel. Three shafts were driven 21m (69ft) into the base of the cliff and filled with a total of some 8,000kg (18,000lb) of gunpowder. The charges were linked by 300m (984ft) of wire to batteries in a shed on the clifftop. Those who witnessed the ensuing explosion described the effect as like 'lava flowing from the side of a mountain'. In typically Victorian fashion, 'splendid, beautiful' were supposedly the first words on the lips of bystanders, as the cliff face crumbled in the name of progress.

The first attempt to dig a tunnel to France was made at Shakespeare Cliff in August 1880. It took the engineers less than a year to realise the magnitude of the task and work stopped after only 830m (2,300ft) of seabed had been excavated. After various other false starts, all tunnelling had ceased by August 1882, but they were not finished with exploiting the potential of Shakespeare Cliff. By the end of the century, coal measures had been struck at a depth of 677m (2,222ft), and its white chalk walls were stained black by the smoke from the colliery at the foot of the cliff.

The march of progress took its obvious toll on the ravens – the last pair was recorded breeding in Kent in 1890 when Queen Victoria was still on the throne. So, when Paul Holt heard that unmistakable 'kronk' of a raven's call over Samphire Hoe in 2005 he dismissed it as a 'silly impossibility'. But he kept looking. A year later he first saw a raven displaying over Shakespeare Cliff in glorious aerial manoeuvre. 'I thought fantastic, they've come back,' he told me in his broad cockney accent, still grinning at the memory.

When I eventually meet up with Paul early on Saturday morning, it is the day after terrorists have laid siege to the Bataclan concert hall in Paris. I have driven along the M20 listening to the death toll slowly being totted up since dawn, and arrive in a state of nervous shock that in recent years has become an all too familiar feeling.

On arrival, I wait in the car for a few moments with the engine turned off to compose myself. The raven's return has come at a time when society seems to be an increasingly cruel place. In that sense, it is a fitting emblem for modern times. In literature and myth, we have often projected on to the raven the most vicious human impulses. And now, as the bird returns, we seem to be regressing to a barbarism from a different age. I look out at the thick sea frets obscuring the view towards France. Someone on the radio is now saying there have been more than 100 innocent people killed.

★ ★ ★

Paul Holt is exactly the sort of person one would wish to meet in the wake of such a tragedy. In his 40s, but with the enthusiasm of a man half his age, he is tall and weather-tanned, with a ponytail crammed under a striped, knitted hat that he has borrowed off his wife the day we meet. He has a bright smile and kind, creased eyes. In the ramshackle hut on the edge of the Hoe where he works, he brews me a cup of tea and begins to tell me about his love for ravens.

Paul grew up in Croydon, on London's southern fringes, about as far from ravens as it was possible to be, and says he always associated the birds with the notion of somewhere else far away from dreary city suburbs – places distant and wild.

On a school sixth form trip to Russia when the Iron Curtain was still up, he watched ravens skirting the snow-covered bleak concrete jungle of outer Moscow. The bird also stalked him in his youth, when he was working and living alone one winter as a conservation officer in the Yorkshire Dales. The prevalence of ravens had ushered in a particularly long and brutal winter. At the time, a series of ghost stories was playing on Radio 4, and he tells me he became too terrified to listen at night.

As we walk across Samphire Hoe, we skirt the foot of Shakespeare Cliff, where great lumps of chalk smashed into fragments across the ground show what can happen if one ventures too close. Rock pipits scatter and re-form around us, feasting on kelp flies that gather when the waves draw back. Stonechats and black redstarts perch on the concrete sea wall chittering in defiance of a cruel wind that has kept most walkers away. Those birds make a beautiful sight, but the whole time our gaze is always shifting upwards, waiting for that black outline on the sheer white cliffs above.

Paul calls himself a 'bird listener rather than a bird watcher' and can do an incredibly accurate impression of a raven's most distinct call, the 'kronk', one of 80 differing voices the species possesses. Ravens can also mimic our own languages. Those with captive birds can teach them an impressive human vocabulary.

The ravens are yet to breed again on Shakespeare Cliff (their nest is a few miles away along the White Cliffs and they come here instead to feed and forage) but this year – and the last – they have been seen over Samphire Hoe at least once a month. Paul has studiously recorded each time. Recently, he tells me, a family of five were displaying over the cliff.

As we walk Paul regales me with stories of his experience of ravens, their 'eternal battles' with peregrines and their speed and agility in flight. He tells me that on one occasion he has witnessed the birds flying upside down. Watching them, he says, one feels as if the birds are watching you back. This is the mirror that the raven holds up to humans.

Our surprise at their intellect has been a feature of our interaction with them through the ages. The Roman naturalist Pliny the Elder described ravens dropping pebbles into a narrow-necked vase to make the water inside rise to the top. In the 1940s, the pioneering Swiss zoologist Adolf Portmann measured the brains of hundreds of birds and concluded that corvids have larger brains relative to their body size than any other group of birds. We collectively refer to ravens as a 'conspiracy' because we wonder what they think and say about us.

On clearer days, the Hoe receives a steady trickle of day-trippers – most of who simply come to walk their dogs, and have little interest in birds. When they see the ravens, their mouths drop, as if they are looking at a creature from another age. Even as they proliferate along the White Cliffs, Paul assures me, grinning widely, that it will not stop there. 'They are going inland as well.'

This book is as much about people like Paul Holt as it is the ravens themselves. The birds are so connected to humans that it would be impossible to document their comeback without dwelling on those who have made it possible, and over the past 10 years recorded every sound and sighting. For the people that follow ravens, there is something about the bird that imbues them. They have spent years in the wilds of Britain monitoring this great exile and what they have discovered is vital.

In order to fully understand the development of this island nation and its people, one must comprehend the history of the bird that embodies our best and worst impulses and symbolises our deepest fears. We have driven them to our boundaries. And now, from each and every outpost of the British Isles, the ravens are coming.

CHAPTER TWO
Bird of Omen

Have you ever wondered why so many places in this country are named after ravens? Just type 'raven' into Google Maps for your local area and you will see the number of options that come up. From Sheffield, where I am writing this now, even the most cursory search reveals six different street names (and a tattoo parlour) whose etymology links back to the raven. I once nearly bought a house on Raven Road, just half a mile away, because I so liked the sound of it.

Or look at any Ordnance Survey map of the countryside, and you will find any number of raven crags, tors, dales, stones, becks, ditches and peaks. As well as these ancient landmarks, cartographers have logged 62 raven place names and 709 road names in Britain. A few centuries ago, before we began to drive the bird further into the wild, most villages and towns would have an unofficial 'raven tree', like the one described in Selborne by Reverend Gilbert White. It was usually the highest and most inaccessible tree in the area, and adult pairs would return year after year to lay their eggs, and local youths would do all they could to steal them.

The raven, more than any other bird or animal, has left an indelible mark upon our landscape. In 2002, an academic called P. G. Moore conducted a biogeographic examination of place names in Britain that derived from the raven. It is not as simple an exercise as it might sound, for even the name 'raven' is only the current word we use for a bird that has been with us since the beginning of civilisation.

According to Moore, English place names beginning with 'Ram' or 'Ran', may disguise a raven connection; so too in the Borders and into Scotland where I've heard 'Corb' or 'Corbie' still used in parts of the Highlands today. In Orkney and Shetland, 'Riven' and 'Ramna' reflects their Viking heritage and its continued imprint on language. In Anglo-Saxon, there are numerous other variations: *hrafn, hrefn, hraefn, hremn, hremm* or *hremu*.

In total, Moore discovered 21 English parishes with names incorporating 'raven' and more than 400 British place names. That total fails to take into account the many other settlements where the raven's influence is less obvious but equally telling. The Isle of Man, for example, even has a raven alongside a peregrine falcon upon its coat of arms, in a nod to its Viking invaders.

Curiously, while we have named so much of our landscape after the bird, raven-related surnames are relatively rare in the British population, perhaps because of our deep-rooted suspicions. However scarce, Moore's research shows there are still among us the likes of Raven, Ravenall, Ravendale, Ravenel, Ravener, Ravenfirth, Ravenhill, Ravenor, Ravenscraft, Ravenscroft, Ravensdon, Ravenspur and Ravensworth. According to Moore, the Scottish patronymic 'McCormick' translates to 'son of raven son'. The first name Ralph is also said to derive from the raven. A dear friend of mine recently gave birth to her first child. She chose Raven as the middle name.

As a nation, we are currently as conflicted as I can remember about what it means to be British, but to me, so much of the modern attempt at definition abjectly fails to understand that our concept of nationality is made up of countless interwoven strands of settlers and occupiers. We are not simply English,

Welsh, Scottish or Northern Irish but the product of the civilisations that have risen up and fallen behind us.

From the coasts of Viking Orkney to Celtic forts; from Anglo-Saxon farmsteads to Roman settlements; from Norman villages to medieval towns, you appreciate the central importance the raven has played in every one of these cultures that make up our own modern identity. The bird, I believe, is what has united us through history, the needle that stitches together the tapestry of this nation and our own sense of self.

* * *

There is a picturesque village in the north of Hertfordshire where something unusual strikes you almost as soon as you arrive. Opposite the thirteenth-century church dedicated to St Faith and a former wishing well, where pilgrims once dropped objects into the water in the hope that they may float and their prayers are answered, stands the old vicarage. On its fence post, cast in resin, two giant ravens stand guard.

Head further into Hexton, and you will find another model raven, looking down in a stately pose from the sign of the village pub (also called The Raven). Behind it is the pavilion of Hexton Cricket Club, formed in 1900, whose players have also adopted the emblem of the king of the corvids – open-mouthed and cawing – on to their traditional whites.

Hexton is an ancient part of the country. Close by is the Icknield Way, said to be the oldest road in England and formed from a series of Neolithic tracks that today extends 177km (110 miles) from Ivinghoe Beacon to Knettishall Heath in Norfolk. The reason why the raven has become the unofficial mascot of the village is apparent a mile or so to the south-west, where the site of an Iron Age hillfort assumes an imposing position. It is known as Ravensburgh Castle and sprawls over some 6.5 acres, making it the largest of its kind in the Chilterns.

Ravensburgh is not an easy place to reach. The fort stands on private land belonging to the owners of nearby Hexton Manor and for much of the year is closed for pheasant shooting. In order to get there, one must walk up a grass clearing carved through the forest, overlooked by hunting seats erected high up on either side and stalked by sparrowhawks gliding silently through the trees, grown fat and bold on the regular supply of meat.

The fort itself, which was built around 400 BC, was designed with protection in mind. It features a single steep slope from the top of the rampart to the bottom of the ditch some 15m (50ft) and supported by what was an additional timber or stone palisade wall. That is the practical reason, but it has also been suggested that Bronze and Iron Age settlers chose high places to be closer to the gods of the sky, whom they revered as much as underworld deities. Here the elements were sharpened and daily rhythms more pronounced – rendering those who lived there physically and spiritually closer to the spirits and helping in turn to pacify their wrath. Either way, nowadays it is a breathless scramble to the top.

Ravensburgh Castle was discovered by local archaeologist and historian James Dyer in the 1960s. He excavated part of the site, but failed to conduct a topographical survey, nor did he ever write up his notes. Only in recent years did Dyer decide to approach the Hillfort Study Group based at Oxford University, to ask for help formalising his work. A Professor of Archaeology, called Ian Brown, agreed to assist but in October 2013 at the age of 79, James Dyer passed away. Professor Brown, who is the author of *Beacons in the Landscape* – a book about the hillforts of England and Wales, has been continuing his work ever since, sifting through the reams of old paperwork and artefacts excavated in the original dig, that Dyer left behind. There are also numerous bones that have been discovered, which are still to be analysed.

It had been assumed that Ravensburgh was the fort of a fearsome Iron Age warlord, but Ian Brown believes it far more likely to have belonged to a group of sheep farmers, with its imposing defences as much to protect their livestock from the wolves that roamed the Chilterns, as invading armies. He has not yet been able to establish the exact reason for the name Ravensburgh, which can be roughly translated as 'town of the ravens' ('burg' denotes a large settlement), but the answer may still be lying in the earth beneath his feet.

In 1969, another archaeologist called Barry Cunliffe – who was a supervisor to Ian Brown when he was still a student – began excavating a different hillfort called Danebury, built in Hampshire in the sixth century BC and used until about 100 BC. The dig has continued almost ever since, and in the process, Cunliffe has made an extraordinary discovery. Deep in the ground, in a similar chalk landscape to the one Ravensburgh Castle stands on, are pits that had been dug and deliberately sealed.

Currently, 2,400 of the pits have been discovered. Largely, they were used for the storage of grain, but many have also been found to contain strata of skeletal remains. Bird bones were recovered from 12 per cent of all the pits, and in particular, those belonging to ravens. Out of 1,200 identified bird bones, more than 70 per cent have been identified as raven skeletons, and 10 per cent crows or rooks.

At first, it was presumed the bones were the inevitable result of ravens commingling with humans (ravens being regular scavengers in Iron Age settlements, which they saw as easy sources of food). Recent studies, however, including one in 2011 by archaeologists Dale Serjeantson and James Morris entitled *Ravens and Crows in Iron Age and Roman Britain*, suggests the raven, in fact, played a pivotal role in ritual and sacrifice.

The authors re-examined the Danebury raven skeletons discovered in the pits and found three sets of wings showing ancient breaks where they had been deliberately snapped off. Other raven skeletons they have examined appear to have cut marks on the wing bone that are strikingly different from traces of butchery and breakage associated with human consumption or the sort of disarticulation that would suggest they had been gnawed by dogs. Instead, the birds had seemingly been slowly and deliberately killed.

Some raven skeletons were found at the base of the pits, and two burials were associated with significant pottery and other artefacts including an iron strip, a quern (a stone tool used for grinding by hand) and a sling stone. Five of 17 raven skeletons discovered had been deliberately buried alongside human remains.

The authors point to at least 11 other Iron Age sites – hillforts, forts and settlements – where ravens and crows have also been buried: Balksbury, Boscombe Down, Budbury, Cow Down, Little Somborne, Owslebury, Rooksdown, Silchester, Skeleton Green, Winklebury and Wittenham Clumps. They also found records of 35 examples of corvid skeletons on Roman sites, 24 of which were ravens. Most of the skeletons were buried in pits, wells and shafts ranging from 3m (9ft) to 30m (98ft) deep. One of the raven skeletons at Winklebury was discovered fully intact and spread-eagled at the bottom of a pit next to pieces of human skull. That skeleton remains on display at the Museum of the Iron Age in Andover, its bones furred by the centuries and its wings spreadout like an archangel.

Why did our ancestors choose to be buried alongside ravens? The theory now being suggested by a growing number of academics and archaeologists is that by placing ravens in these pits, they were offered up as gifts to the gods of the underworld. Before being buried, the feathers were also plucked from the birds' bodies and used to decorate weapons and armour to terrify and intimidate their enemies. In a society where birds and animals were seen as a continuum of human life rather than something existing alongside it, the ravens were there to perpetuate the existence of the human soul and be our companion and guide in the afterlife.

* * *

One December evening a padded envelope slipped through my door containing a letter and a CD. The note was crafted with rare attention in beautifully neat fountain pen ink and wished me a happy Winter Solstice. It was written by a man I know called Michael George Gibson, a family friend, gardener, husbandman, and expert on the oldest written poetry this country has ever produced.

The CD was a translation and reading Michael had undertaken of an Old English poem first written down in the tenth century. It was found in *The Exeter Book* and presented to Exeter Cathedral by its first bishop, Leofric. Originally untitled, it came to be known as *The Wanderer.* The poem, Michael said in his accompanying letter, is an elegy with an emphasis on exile; the song of a warrior from the heroic age of European migration who is bereft of lord, kinsman and companions, and laments the loss of former joys.

In the recording, he said he had closely followed the alternative four-beat pattern of the lines, and used only modern English words in his translation that have their roots in Old English. He had also added the deep kronks of a raven, the perfect melancholic accompaniment to this ballad of loss.

For some 15 years in the 1980s and 1990s, Michael lived almost self-sufficiently off five acres of Cheshire land. He slept in farm outbuildings and later a wooden caravan without electricity, and kept goats, bees and a small herd of St Kilda's Hebridean sheep, whom he sheared by hand. He made his own goats cheese, grew his own vegetables and trimmed the meadow grass with a scythe to use the cuttings as his bedding. His only source of heating was a paraffin lamp over his desk and the evening bonfires he lit outside each evening, fed with birch, beech, hawthorn and apple. Here by the flickering light, he taught himself ancient poetry. 'It was a hard but a sweet life', is how he remembers those days.

I listen to the 115 lines of *The Wanderer* read in both modern and Old English and accompanied by the sounds of the raven and Michael's tin whistle. The raven, he tells me when I later get in touch, appears regularly in Old English poetry. A month or so later he is travelling down to Oxford to attend a lecture by Simon Armitage and a folk gig by his friend Dr Mark Atherton, who teaches Old English at the university and plays the Irish bouzouki. Michael invites me to come along and hear more about the place of the raven in our ancient literature.

When we meet in a pub after the lecture, Michael is wearing a sprig of holly in his hat intended as a symbol of non-vocal protest against the perceived inaccuracies of Armitage's translation of the fourteenth-century poem *Gawain and the Green Knight*. He implores me not to get him started on that for fear he will never stop. Instead, we get talking about the shadow of the raven that looms over our history.

Michael tells me his attempt to live a more self-sufficient way of life was all about better understanding the essence of our language. The pursuit of simplicity and daily chores required to sustain such an existence helped him to dive down into what he calls 'the rhythm' of Old English verse. Either tilling the soil, chopping logs or, on particularly cold nights, running around the lanes of Mobberley to get the blood moving through his body, this ancient beat informed his work and seeped into his bones.

I find this notion of the rhythm of language and nationhood that Michael discusses articulates a part of my own work that I have been trying to achieve; the idea that by acknowledging the wilder part of ourselves we gain a better understanding of our identity. That is what the raven has always represented in this country: a more visceral manifestation of the human condition. The rhythms of the bird, the beats of its wings and calls that echo out over a landscape with metronomic clarity, interspersed with our own.

In our earliest texts, the raven features prominently in the so-called 'beasts of battle' motif, alongside the wolf, eagle and boar, as a harbinger of violence and bloodshed. In the Old English poem, *The Battle of Brunanburh*, for example, the bird is mentioned as the 'dark-plumaged, horny-beaked black raven', while in the Anglo-Saxon poem *Judith*, it is: 'the corpse-greedy bird, the black-coated raven'. In *Beowulf*, the longest epic poem in Old English, nearly every mention of the raven is to do with death and destruction. There is a slaughter in Hrefnesholt that translates as the 'Ravenswood', and on two other occasions, the birds are referred to as corpse-eating ravens. Curiously in spite of that, the very first mention in *Beowulf* describes the raven as 'blithe-hearted' and 'calling out with the rising sun to proclaim the joy of the gods'. This suggests a far more complex relationship with the raven than we might initially imagine.

A few weeks after I met him at his gig with Michael, Mark Atherton emails me a few of his own findings that further hints at the magical properties of the raven beyond mere doom-monger. In the Old English herbarium, a compilation of the herbal remedies of Anglo-Saxon England, he has found mention of Raven's foot, a plant recommended to treat dropsy

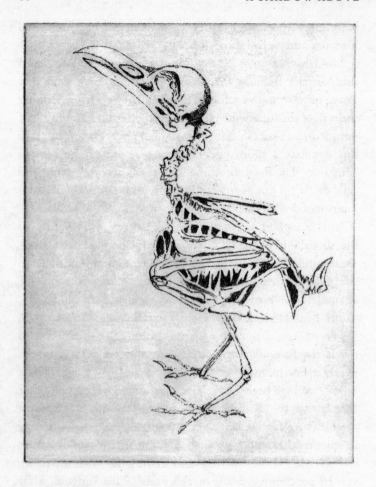

(swollen tissue in the body), and an orchid known as Raven's leek to use on severe wounds.

He has also discovered several mentions of ravens in the stories of the saints. St Guthlac, for example, the seventh-century son of a Mercian nobleman, was initially tormented by ravens who stole his gloves and letters but eventually returned in an act of contrition to give him lard to protect his

feet from the damp. A raven guarded the dismembered hand
of St Hugh of Lincoln and carried a loaf of bread each day to
St Paul the Hermit. Such stories show a much more ambivalent
view of the traditional one-dimensional summary of
the raven as a simple symbol of ill omen and hints at why
those who lived and died in England's Iron Age forts may
have wished to be buried alongside them.

* * *

Sometimes we name places after ravens and sometimes they
seek them out by their own volition. So it was a few years ago
on a church spire in Penarth, in the Vale of Glamorgan,
where a pair decided to build a nest. They have been there
ever since, as adult raven pairs are known to return to the
same site year after year to rear a brood. In a joyful coincidence,
across the road lives the country's leading expert on animals
in ancient life and myth: Professor Miranda Aldhouse-Green.
'The cry is very melodic,' she says. 'And very human. We
coexist very happily.'

I want to speak to Miranda after reading one of her early
books, *Animals in Celtic Life and Myth*. In it, she makes
fascinating discoveries of ravens being seen as 'messengers of
the underworld', as well as symbols of the 'pitilessness and
carnage of war' in Ancient Britain and the multitude of
cultures that comprised Ancient Britain. As with the Anglo-
Saxons, in Celtic belief, ravens are not merely seen as birds of
the battlefield, but also prophets with the ability to shift
shapes between different worlds. The only other bird similarly
revered by the Irish Druids was the humble wren, our second
smallest and most common breeder which, while bombastic
in voice, possesses little of the raven's stature.

In her book, Miranda tells the old vernacular tales of
Ireland, where the goddesses of war, the Morrigan and the

Badb Catha, could change at whim from human to raven form, squawking dreadful omens and terrifying armies at their presence. The belief that ravens could predict the outcome of a battle was a powerful force that ran through ancient armies. The Celtic raven god of war, Lugh, was warned by raven familiars of the approach of enemies. As with the Norse god Odin, the Ulster hero Cú Chulainn kept two magic ravens as oracles. In the old Welsh myths, written out in the four branches of *The Mabinogion*, ravens also occur prominently. In *The Dream of Rhonabwy*, Owain has an army of ravens who possess magical powers of recovery.

Miranda tells me there are both British and Gaulish coins depicting the image of a horse bearing an enormous carrion bird holding the reins. She also writes of a dry well associated with the Romano-Celtic temple at Jordan Hill in Dorset that was filled with pairs of tiles (16 in all), inside each of which was a coin, and the skeleton of a raven.

Still, the main iconography of the raven appears on the battlefield. A helmet, dating back to the fourth century BC, was discovered in Celtic Romania. Shaped like a raven, it was decorated with coral and red enamel and featured deliberately articulated wings designed to creak up and down as the warrior charged toward his enemies. There were also Iron Age war trumpets known as carnyx; described by Caesar as hellish instruments of war and taller than a person, the mouthpiece was fashioned in the form of a wild boar, wolf, or occasionally, a raven. The massed trumpets could produce a terrifying noise. Sight, sound and the imagery of the 'beasts of battle' were all part of psychological warfare and a matter of life and death.

'Of all the individual bird species in the written mythology, the raven is perhaps the most complex and interesting,' Miranda writes in her book. 'The blackness, cruel, tearing beak, pitiless eyes and its predilection for dead flesh, endowed the raven with this sinister imagery. Only occasionally is the raven projected in a more positive light, as friend to man, appearing to warn and to protect.'

Miranda Aldhouse-Green is quite different to the clichéd perception of a professor of archaeology. When we speak, she has recently returned from a monastic retreat and has just bought a brand-new Mazda sports car. Her interest in the British Iron Age and Celtic world – and the raven by proxy – was formed after initially studying Roman history at university, but quickly focused on the ancient cultures obliterated by the all-conquering Imperial army. She has, she admits, always felt a propensity to defend the underdog.

The veneration of animals in our traditional cultures is, Miranda believes, as telling an indicator of any as to the motivations of our ancestors. The raven has always been lurking in her thoughts, but the pair now nesting on the church steeple outside has recently prompted her to begin reflecting more deeply on the role they have long assumed – and how they might be interpreted in the modern age.

'It seems to be a very strange and unstable relationship that we share with ravens,' she says. 'The fundamental thing about ravens is they like humans, so they stay close. It's an almost symbiotic relationship and something very fundamental that goes back to the first human settlements and domestication of beasts.

'I wouldn't be surprised if people didn't make links and connections between the fact ravens are coming back at the same time as we have more instability in terms of the world than we've known for centuries. The return of such an iconic beast at a time of deep unease means connections will be made. Doom-mongers will say it's a beast of death. I would tend to be much more positive about the raven. The recolonisation of any wild species is something to be celebrated.'

By the time of the Roman occupation of Britain, the transcendent role of the raven became cemented in our culture. The raven was considered by the Romans as the most prominent of all the birds of omen, or oscines, as they were known. Mithraism, a religion that centred around the god Mithras and persisted for centuries in the late Roman Empire, relied upon the raven as one of its most sacred symbols. When the Temple of Mithras was discovered in Walbrook, London, during a construction project in 1954, the wing bone of a raven was among the relics identified by archaeologists.

The Romans, like the Celts, were in thrall to the raven's capacity for violence and destruction. 'Fling him to the ravens' was a typical curse of the Empire. The raven's beak, sharp enough to tear the toughest flesh from a bone, proved a particular object of fascination. The author Reginald Smith in his *Bird Life and Bird Lore* mentions that the Roman champion Valerius, whose helmet bore the beak and wings of a raven, was given the honorary nickname of Corvus for his bravery in battle. During the Punic Wars between Rome and Carthage, Admiral Gaius Duilius invented a grappling hook with a vicious spike on the end to storm Carthaginian ships. He called it the Corvus or Korax, after the raven's beak.

Ravens were also revered for their intelligence and the assumption that they bore messages from other worlds. Pliny the Elder wrote of a raven pair who, during the reign of Tiberius, nested at the top of the temple of Castor and Pollux in the Roman Forum. One of the young birds learnt to say 'good morning' but was killed by a cobbler who believed it was bringing his rival businesses good luck. A furious mob rounded upon the cobbler and killed him for the murder of such a sacred bird, and a public funeral was afforded the raven. Its body was laid upon a bier borne aloft on the shoulders of two Nubian slaves and decorated with flowers. A trumpeter led the procession to the funeral pyre, where the raven was torched to allow its soul to ascend to the gods.

The idea that we somehow turn to this savage soothsayer for guidance persisted with the fall of the Roman Empire and the Christianisation of Britain. As churches were built, often in the exact same place as Neolithic sites of pagan worship, the bird retained its totemic status. As with the old stories of the saints in the Bible, the raven is afforded a special status far beyond battlefield scavenger.

These are the often-conflicting strands that make up our ambivalent modern view of the raven; the imprints in our collective memory each version of civilisation leaves behind. Most non-bird enthusiasts I have spoken to about my book do not even necessarily know what a raven looks like, and yet they recoil from instinct. When I am at Hexton by the raven statues on the old vicarage, I cannot resist ringing the buzzer on the electric gate and chatting to the very pleasant man who comes out to greet me. It turns out they were a commission from a sculptor friend, designed in honour of the name of the Iron Age fort that looms above his house. 'Evil bloody things,' he says of the raven before checking himself. 'Actually, I've never met one before.'

During my research, I come across another book in the British Library, written by the Reverend G. Oliver in 1866. Entitled *Ye Byrde of Gryme*, it is intended as a historical and topographical survey of Grimsby, but the author prefers to tell his story through the guise of an imaginary raven. Oliver never quite fully explains why he introduces the raven, which appears to him while dozing under a pear tree at the start of the book, apart from he regards it as a vehicle of history. The bird of his creation arrived at the mouth of the River Humber in 645 AD, when a real-life Viking port called Ravensrodd used to exist, but has long been washed away by the shifting tides. His raven, Oliver claims, has seen 'the Briton, Greek, Roman, Saxon, Dane, Norman and all other successors in this country.'

'My raven is a personification of memory and the depository of things past, present and to come,' he writes in his author's notes. 'Do you approve of the plan or do you doubt the prescient qualifications of my informant? Well, that is simply a matter of taste.'

It is fascinating to trace this line from the Celtic myths and Old English texts to modern literature and find the raven's role in the narrative almost unchanged. What is it about these spiritual abilities and historical ambiguity that we invest in the raven? How does this notion persist that the bird can tell us more about our world than we are able to ourselves?

* * *

I spot sparrowhawks, buzzards and a lone, shrieking red kite stalking the hills around Ravensburgh Castle, but despite straining my ears and keeping a constant eye on the sky, I never catch sight of a raven. They are back in this part of Hertfordshire and advancing all the time, though not yet abundant enough to return to this former citadel. Still, all over the country, ravens are now once more taking up residence in the places that have borne their name for centuries.

'Rewilding' is how they term it in modern parlance, and the beasts of battle of Anglo-Saxon poetry are now slowly living back among us. The wild boar population in the Forest of Dean is growing all the time, while the white-tailed sea eagle was reintroduced to the Isle of Rum in 1975, and has now spread along the west coast of Scotland. At the time of writing, there are 500 breeding pairs of golden eagles in

Scotland, even if England's last golden eagle went 'missing' (as is sadly often the wont with British birds of prey) in 2016 and is feared dead. A wolf cub pack is currently being monitored at Wildwood Escot Park in Devon with a tentative view to introducing the animal back into northern Scotland in the near distant future, although the recent furore an escaped lynx on Dartmoor prompted, during three precious weeks of freedom, shows how far off many are from acceptance. And then, of course, there is the raven.

Mercifully, there are no battles anymore, or at least not the sort where carrion birds can pick over the corpses of the fallen. But these animals, to which we attach so much meaning and significance, are returning at a moment when our version of civilisation is beginning to flicker and dim.

We are, it is said, in the midst of a new geological epoch: the Anthropocene, bringing an end to the Holocene, which marked the past 12,000 years of relatively stable climate since the end of the last Ice Age. Scientists argue the Anthropocene began with the end of the Second World War and the mass transformation of the landscape for food production. It is marked by carbon emissions, sea level rise and the global mass extinction of species – the sixth in the entire history of the earth. Extinction rates are now running at 100 times their natural level because of deforestation, hunting, pollution, overfishing and climate change. Thousands of species of plants and animals are disappearing before our eyes.

It is the scavengers that will flourish in this new era, those accustomed to feeding off the waste that humans leave behind. Yet the raven's return can also inspire hope. If this bird, which in years gone by we have done our level best to wipe off the face of the earth, can be back living among us in places where it has not been seen for centuries, then there is hope for other species too. In this sense it is not yet possible to say what the raven will come to represent for our own age: the vanguard for a new return of the wild into human settlement, or the destroyer picking over the bones of a ruined world?

On a wet, misty spring morning, I drive out through the Peak District to Ravensdale. My route takes me through the gritstone faces and sweeping heather moorland of the Dark Peak and into the steep limestone crevices of the White Peak. Ravens returned to nest in Derbyshire in 1992 after being absent from this landscape for nearly a century, eradicated from their former grounds by gamekeepers who deemed the birds a pest. Slowly, they are re-establishing themselves across the Peak District despite ongoing cases of persecution. In 2005, the RSPB published a 10-year study of 18 raven pairs in Derwentdale. They had fledged only on average 1.8 young per nest each year, the lowest figure for anywhere in the country. In eastern Derwentdale an entire nest and its clutch of eggs had been removed from a cliff, while a second raven nest and its young chicks had also disappeared. The study pointed to the driven grouse shooting interests across the moors and appeared to lay the blame squarely at the feet of gamekeepers who fear ravens predating upon grouse nests.

With its sweeping uplands and steep, sparse valleys dotted with human settlement, this is raven country, and even bullets and surreptitiously-placed poison have not been enough to prevent the inexorable return of the bird. In 2013, after years of waiting, a raven pair returned to nest at Ravensdale. They have been here ever since.

I drive up alongside the River Wye to the site of the old Cressbrook cotton mill, as grey and foreboding as the sky, and then turn down the one-track road to Ravensdale. There is evidence of human habitation here dating back to the Bronze Age, but the current settlement takes the form of two terraces of six millworkers' cottages, built in 1823.

Surrounded by towering limestone crags and deep woods, Ravensdale is a solemn, isolated place. Despite being well built in local stone, with slate roofs and gothic windows, the cottages were not popular with millworkers and became known locally as 'The Wick' shortened from the full nickname 'Bury-me-quick'. I have seen one photo of the village and its surrounding allotments (now long gone) from the nineteenth century, which shows smoke pouring out of each chimney stack, and the foreboding hills looming above. The highest peak is Raven's Crag, which juts out over the valley like a headstone.

It is early spring, and raven chicks will be starting to emerge from their nest. I park up by the cottages in the middle of the woods. I'm surrounded by beech trees, whose trunks have turned russet with age, and carpets of bluebells and flowering wild garlic, the heady scent made pungent by the light rainfall.

When the birds returned in 2013, the crag was closed to climbers during the breeding season, and a sign stuck to a nearby stone wall informs that the ban remains in place. As I look up through my binoculars to locate the raven nest, a man walks up next to me with a cigarette between his teeth and lights it with a blowtorch. He introduces himself as Mike, a builder, who only lives two miles away and is often called out to work on these cottages, most of which are used as holiday homes. He loves this time of year, he tells me, because it is often just him and the ravens in Ravensdale.

Almost as soon as he mentions the word, one of the adults takes off from the nest, a huge, sprawling construction as intricate and imposing as the stonework on the mill down the valley, and soars up high into the air calling out as it goes. The raven kronks, twists, and glides in and out of the murk like a shark's fin slicing through the waves. A few jackdaws rise to meet it, and the raven simply shrugs them away. The bird performs a couple of long, lazy loops over its domain. Its deep, guttural refrain echoes down the valley, over the wind-rushed trees, smattering rain and sated river racing fast over the rocks. Its partner and young are nowhere to be seen, presumably foraging or practising their flight over the brow of the cliff. This is a simple performance, an assertion that Ravensdale is once more occupied territory.

CHAPTER THREE

Ravens and the City

It is a winter's evening and close to dusk. I am standing alone on Bristol's Clifton Suspension Bridge, not far from a group of teenagers on a bench, sparking rolled cigarettes and playing music videos on their phones. They are huddled into their hoods against a biting westerly wind, which sings through the wrought iron chains and suspension rods of the bridge. Behind me, the toll gate jerks open and shut, waving to the passing cars.

I have my binoculars up, scanning the Leigh Woods and Avon Gorge. A jay breaks underneath the bridge and I follow its chestnut flash, darting between trees, over the muddy, churning River Avon, whose currents twist and spiral 75m (245ft) below. 'Don't jump,' one of the teenagers calls out to me in broad Bristolian sending the rest into fits of giggles. True enough, on the old Pennant Stone piers of Brunel's 1864 bridge are black and white signs urging the desperate who have ended up there, to call the Samaritans.

I wander on to the east bank and up Clifton Down along the Avon Gorge. It is a Wednesday evening and joggers in bright lycra pound past me along the paths. Their headphones are plugged in and eyes fixed firmly ahead as they run off the stresses of a working day. Among the brambles and trees closer to the edge of the gorge, I spot glimpses of tents where homeless people have found a patch of city scrub to sleep. I count three different tents in only 100m (328ft) along the Downs. Across the gorge, among the whitebeam of Leigh Woods, I see another blue tent pitched in a small clearing.

Eventually I reach a viewing point on the edge of the cliff. On the sheer slopes of the gorge, graffiti artists have sprayed in giant letters the name of a pirate radio station and added their own tags and self-portraits to the limestone faults. Far below me, rush hour traffic snakes past alongside the river, inching forward like a glow-worm through the dusk.

As I stand watching, I am joined by an elderly, well-spoken man in a cashmere scarf and wide-brimmed hat. He tells me, he walks up here each evening to catch a glimpse of the peregrine falcons and (now) ravens that nest on these cliffs. We stand together chatting for a while, sometimes making eye contact, often looking beyond one another in that strange, impersonal manner birdwatchers do; bridging the human and avian worlds with one boot in each.

They say the mornings are the best time to see birds, and ravens, like other corvids, are notable early risers. But there is something about these evening sightings that my companion and I agree, give a particular thrill – a momentary pleasure bathed in fleeting sunset hues. I will always remember watching an osprey planing over the water in Talisker Bay on the Isle of Skye, from the rocky beach where my wife and I had pitched our tent for the evening. We stared as it sailed, wings cocked razor-sharp across the sinking orange sky, sending crows scattering up like embers, before disappearing into the gathering night.

As we are speaking, out of the corner of my eye I see a raven take off. It arcs down at rocket-speed from the top of the bank and into the woodland across the river and out of sight. The raven flies more like a peregrine than a bird that is commonly known for soaring over these cliffs. Only its giant wingspan tells it apart from this distance. The raven is just airborne for what feels like a few seconds, time for my companion to spin around, and both of us to clap each other on the shoulder celebrating a joyous unspoken experience shared between total strangers.

'The hawk on fire hangs still,' wrote Dylan Thomas in the poem *Over Sir John's Hill* inspired by his dusk walks over the Towy estuary. Our raven, on the other hand, sinks like a stone.

Night upon us, we bid farewell and I walk back along the Downs to the home of my friends Vianet and Helen and their family. It is bath-time and their young son whoops with delight as he makes waves on to the floor, ignoring the scolds of his parents. The wood-burning stove crackles in the living room and pots bubble away on the stove. The house is filled with warmth and love. I think of those sleeping out in their tents on the Downs that January evening, and the harshness and precariousness of survival in the modern city.

* * *

'Why the raven?' they ask me over supper that night, and I end up mumbling a response into my wine glass. It is a question I find I still struggle to answer no matter how many times I am asked. For certain, there are more obviously beautiful birds to follow around the country than this giant king of the corvids. I chose my other favourite birds because they remind me of a sense of place: grey wagtails and dippers for the uplands of Yorkshire, greenfinch for the birds that settle on the firethorn outside my study window each autumn. These birds exist regardless of human activity. The wagtails and dippers dance up fast-flowing moorland streams because of the proliferation of aquatic insects there. The greenfinch comes because the shrub is in fruit. My presence means nothing to them aside from triggering some instinctive threat response.

The raven, though, has a far closer relationship with us. When a raven sees you, it knows you are human and behaves accordingly through the myriad traits that have evolved over thousands of years. Throughout history, humans and ravens have been co-dependents. The ravens upon us for the scavenging opportunities that humans leave in their destructive wake, and us upon the ravens for some sort of spiritual meaning to explain our own motives. It is why the bird has become so ingrained in our culture and replicates so many of what are wrongly perceived to be solely human character traits. Unfashionable as it may seem to some twentieth-century scientists who loathe the idea of viewing the animal world through humanity's prism, the bird is a study in anthropomorphism.

I am fascinated by the raven as much for its soaring aerial displays as what it tells us about ourselves. Perhaps this is what the Mad Hatter was getting at with his riddle in *Alice in Wonderland* that 'a raven is like a writing desk'. Its presence forces us to think and see things differently.

John Marzluff, the US wildlife professor and author of *In the Company of Crows and Ravens*, gives five principal reasons why the birds have evolved so closely alongside man: hunting and gathering, expansion of agriculture, war and aggression, urbanisation and recreation. These have all provided new food sources for the omnivore ravens, which are clever enough to adapt quickly to whatever opportunities come their way. Marzluff argues that it is particularly the increasing urbanisation of human life (by 2030 more than 60 per cent of the global population is expected to live in cities) that is having the greatest impact on ravens in the modern era.

'Our buildings, utility poles and other structures provided new nesting and roosting sites. Our sewage ponds, watered lawns and lakes bring life-giving resources to birds in arid regions. Such subsidies allow many corvids to exploit new environments or attain enlarged populations where they were formally rare.' The proliferation of these anthropogenic nesting sites, he writes, has allowed raven populations to 'increase exponentially'.

In modern Britain, they are now moving back into our cities and urban spaces. When you start looking, you see ravens on motorway bridges, rubbish dumps, power lines and buildings. People still see ravens as birds of the extremities, but that is only because we have forced them there. In fact, ravens are relaxed city dwellers. Centuries ago that is where you would have found them; on the cobbled streets of Britain, sifting through the offal and other waste deposited outside homes.

Why did they leave? Improved sanitation, certainly, and the Victorian systems of water supply and waste treatment, that cleaned up many British cities, would have lessened their urban feeding opportunities. At the same time, increasingly intensive agriculture would have presented greater spoils outside of the city. And then there was the attempt to purge ravens altogether under the Vermin Acts. Once they had started moving in to the countryside, landowners sought to clear their estates from anything that might threaten commerce, be it farming or shooting.

The reasons were varied, but entirely the product of human activity and the raven retreated to the furthest edges of the country. Even if they depended on us for food, our ravens still learned to avoid us. Before Dylan Thomas, the eighteenth-century poet John Dyer also meditated on the wild Towy Valley in his poem *Grongar Hill*. He describes an ancient castle, long abandoned by humans and returned to the wild. 'Tis now the raven's bleak abode.' Such a line reflects the thinking of the time: man and bird were supposed to occupy different places.

Not so in the modern era. In his masterly book *The Raven*, published 20 years ago, the ecologist Derek Ratcliffe mapped the bird's decline and slow subsequent advance across Britain in the latter decades of the twentieth century. When he died in 2005 (on a trip to the Arctic Circle), the obituaries described Ratcliffe as the most influential British naturalist of his generation. He was the first to link the decline of birds of prey with agricultural pesticides, and the architect of the 1977 *A Nature Conservation Review*, which identified the most important wildlife habitats in Britain and became known as the modern Domesday Book of nature.

Ratcliffe wrote about the raven being 'inextricably interwoven with human activity through much of its range'. Thousands of years ago the birds learnt to become camp followers and settlement scavengers and, as a result, he argued, they have always been afforded special significance by people. 'Humans also provide the raven with a food supply more directly,' he wrote. 'When the opportunity arises, the raven finds a meal of defunct human flesh as acceptable as that of any other large animal.'

In his book, Ratcliffe presents a number of maps detailing changes in breeding raven pairs across Britain in the second half of the twentieth century. Between 1968 and 1991 (the period before the raven boom really began) the bird was restricted entirely to the western fringes of Britain. The whole centre, east and north-east of England were almost entirely moribund. Ratcliffe contrasts this map with the prevalence of grouse moors, and the lack of crossover is startling. The raven's failure to establish itself in the marginal lands and foothills of the Pennines, Cheviots, Southern Uplands, and Eastern Highlands (which are ecologically similar to the western parts of the country where the raven was abundant) 'can only readily be explained by assuming an inhibiting degree of persecution', he writes. Ratcliffe also makes mention of a map produced by Norman Moore in 1957 showing the polarised distribution of buzzards and gamekeepers. 'Precisely the same correlation is still found in the case of the raven.'

The story of the raven is also closely connected to another British bird of prey: the red kite. The birds used to forage side-by-side in British cities and, as medieval dustbin crews, shared similar protected status by royal decree. If you killed a kite in medieval cities, you risked capital punishment. Shakespeare described London in *Coriolanus* as the 'city of kites and crows'.

The kite ended up being purged with perhaps even greater vigour than the raven. It was declared extinct in England in 1871 and eight years later the last of the birds were driven out of Scotland. By the 1970s, the British red kite had reached the final bottleneck of extinction, the entire population emanating from just one female in the bird's last citadel in mid-Wales, whose old oak woods gamekeepers and egg collectors could not reach.

At this point, however, the stories of the raven and red kite diverge. The kite was saved in Wales, through what is regarded as the longest-running conservation programme in the world. Residents, farmers, and latterly Gurkhas and SAS commandos, roped in by the RSPB, started guarding nest sites to deter would-be egg collectors. From 1989 a reintroduction programme was launched across Britain, bringing in breeding pairs of red kites from the Continent and rehoming them on British soil. In 2013, there were estimated to be 1,600 breeding pairs in Britain, an 874 per cent increase since 1995.

The raven's return, while similarly successful, has not been the product of any specialist conservation programme. Protections may have been placed upon them – as with other wild birds under the 1981 Wildlife and Countryside

Act – but nobody has reintroduced the raven to places where it had previously been extirpated. The raven has returned to cities like Bristol entirely under its own volition, moving into the spaces we unwittingly make for them.

A few years ago, I visited Gigrin Farm in Powys where the red kites are fed at 3 pm every day. The ritual, for that is what it has now become, is always the same: farmer Chris Powell drives up in a red tractor and heaves down 90lb (14 stone) of gristle-grey offal on to the grass. The birds gather in their hundreds and then take it in turns to hurtle down to feed and veer up with their tail feathers splayed, forming a single swirling vortex from earth to sky. A permanent daily feeding station for the red kites has been in operation here since 1992 and nowadays attracts bird enthusiasts from all over the country to these sparse, steep valleys. Before they started feeding red kites, Powell and his father Eithel used to leave offerings for the ravens too, to keep them away from their sheep.

When the feeding is over, I tell Chris about my interest in ravens and he ushers me over to a locked barn. Inside on the perch is a red kite with a broken wing that has been sent to Gigrin Farm to recuperate. Next to it, staring out through the gloom with gleaming eyes, is the unmistakable black shaggy shape of a raven. Powell tells me the raven is a rescue bird donated to him by a wildlife centre because of fears it was becoming too domesticated, learning to mimic people and no longer associating with itself as a bird at all. The hope is that during this enforced period of isolation, the raven will become used to the wild again before being re-released into the countryside. As a result, we are not allowed to speak within its earshot and can only be around it for a few moments.

I take a few photographs of the red kite standing utterly indifferent on the perch. Even though the flash is turned off the raven still appears curious to know what is going on, hopping over to the kite, so it too appears in camera shot. This lonely bird, caught in an existential crisis, makes a melancholy sight next to the purely instinctive red kite, whose edges have been so sharpened by evolution to hunt, kill and procreate. The kite can be released as soon as it is physically recuperated. The raven, however, requires a period of mental detachment from humans to enable it to survive happily in the wild. It is a fascinating concept that a bird can think deeply. I stare into its eyes, wondering what it makes of the strange man taking photographs on his phone. Powell taps me on the shoulder and beckons me out: it is time to leave the raven to the darkness again.

* * *

The British Trust for Ornithology (BTO) has an updated version of the raven population maps Derek Ratcliffe used 20 years ago. The latest data (recorded between 2008 and 2011) reveals that breeding ravens have spread to all but the eastern fringes of England. The bird has now returned to every county in the country. Slowly, pairs are beginning to reappear in cities, too.

Our western conurbations have been repopulated first, and, as is the natural inclination of the raven, in suitably evocative locations. For a time, ravens were nesting on the Swansea Guildhall clock tower and they have also moved on to Chester's Town Hall. When the nesting site was established in 1996, it was deemed so significant that locals added a raven character to Chester's annual Midsummer Watch Parade – one of Britain's oldest festivals held to celebrate the summer solstice, dating back to 1498.

The ravens since moved their nest to Chester Cathedral in 1997 and reared young among its gargoyles numerous times over the following decade. More recently, ravens have nested among the spires of Cardiff University and, at the time of writing, a pair has settled on the old red brick chimneys of Wigan Town Hall, looking out over a hunting ground that includes Debenhams, WH Smith and Sports Direct.

A pair of ravens first returned to Bristol's Avon Gorge in 1993, and the birds have nested there ever since. According to information kindly passed on to me by the Bristol Ornithological Society, between 1993 and 2012 some 50 active breeding sites were identified across the region. In the 2007–11 BTO Bird Breeding Atlas, ravens were found in 284 of the 400 tetrads (2km by 2km squares) spanning the Avon region. Of the areas mapped, they were found in 93 per cent of the Mendips, 70 per cent of urban sites, and 55 per cent of the Somerset Levels – the lowland floodplains around Weston-super-Mare.

In 2012, it was estimated that there were 100 breeding pairs (or one pair for every four tetrads) in the Avon region. Within the metropolitan area of Bristol, there were seven confirmed nesting sites including Leigh Woods and the old water tower in the Avon Gorge. The 2016 Breeding Bird Survey showed ravens present in 26 per cent of the total squares surveyed – the highest proportion yet. There were three raven pairs recorded in the Clifton Downs area alone.

The birds can now be heard overhead almost anywhere within the city. Over one winter I followed the daily sightings recorded on the Bristol Ornithological Society's Avon Birds Blog and it makes fascinating reading. Ravens are now being spotted in every conceivable urban setting: over Oldbury nuclear power station (a decommissioned old hulk where the Bristol Channel meets the River Severn) and calling from the floodlights of the Bristol county cricket ground. They have been seen stalking the fashionable streets of the city's Montpelier quarter, feeding for muffin scraps dropped by hipsters, and kronking over St Werburghs City Farm. Ravens have also been frequenting city parks and the old B Bond tobacco warehouse built in 1908 to house goods coming into the Bristol Channel from across the British Empire. Nowadays, it is the site of the city's public record office as well as a modern eco-conference centre.

Aside from the dedicated volunteers monitoring these sightings, I wonder how many in the city actually notice the returning ravens as they go about their daily business? In truth that doesn't matter a jot. Even if they have no idea they are back, it makes my heart sing to think of ravens and people coexisting again so happily in an urban setting.

Soon after I first get in touch with the Bristol Ornithological Society, I receive an email from a man called Richard Belson, the secretary of the club. He lives in a town called Nailsea, 10 miles or so from the centre of Bristol, with a population of some 15,000 residents. Richard tells me that 2016 was the first year ravens had returned to Nailsea in the 40 years since he had lived there. A different pair had previously been nesting for some years in the grounds of Tyntesfield, a Victorian Gothic Revival manor house owned by the National Trust, but now apparently, they were spreading.

Generally, Richard spots the Nailsea ravens at around 8.30 am. On one occasion, he told me, one shot past the house with something – possibly an unfortunate blackbird – in its claws and three herring gulls in hot pursuit. A few days later he noticed a raven chasing a sparrowhawk out of the area. I had already heard from a friend of mine who also lives in Nailsea that he was hearing ravens over his house in the mornings. After staying the night at Vianet's in Bristol, we agreed to investigate.

I have known Vianet since we were both students at Leeds University more than a decade ago. We met at a local cinema called the Hyde Park Picture House. He was training to be a projectionist and I was a hapless security guard. Aside from the dreaded Saturday morning shift, when local schoolchildren would terrorise me during the 50p-a-ticket matinee showing, I would spend my days watching free films and thinking I had the greatest job in the world.

Originally from the Republic of Congo and having grown up in a tough part of Paris, Vianet had come to England to study film and worked at the cinema as a volunteer. We would often smoke roll-ups out on the cinema steps together in the evening when the film was playing and Vianet would talk about his dream of becoming a documentary filmmaker and me, a writer. We used to speak a bit about wildlife back then, and the moors that surrounded our city. Little did we realise how this burgeoning interest in nature and human stories would come to shape our working lives.

These days, Vianet is a cameraman and presenter at the BBC's Natural History Unit in Bristol. Still, our old student ways die hard. We drive over from Clifton to Nailsea with the windows of my car rolled down to try and alleviate our hangovers from the numerous bottles of beer and wine we had drunk the previous evening.

I had been told by Richard to head for a place called Nowhere Wood. It is in a curious location, surrounded on all sides by houses and a school. At first, Vianet and I wander up and down the main drag unable to find it. A local man stops to ask us what we are looking for and his eyes light up when I mention the ravens. He has seen the birds for the first time himself, he says, and he leads us down the road and on to a footpath that heads into the wood.

The only reason this wood still exists and was not gobbled up long ago by developers was because of industry. Trendlewood, to give its official name (locals decided on the nickname because of Nowhere Lane that runs through it) was, until less than a century ago, home to several Pennant

Sandstone quarries and the air would have been thick with dust and grit as quarrymen cleaved the rocks apart. This particular type of sandstone was used in the twentieth century for paving and new-build homes. But the money eventually bottomed out and around 90 years ago, the last of the quarries closed. Since then, trees including sycamore, ash and hazel, and a magnificent array of ferns have grown over the spoil heaps and ivy shrouds the hard-worn edges. As we walk through the wood, we spot dunnocks, robins and blue tits and hear the distant thrum of a woodpecker on a branch. Nature reasserting itself.

Our companion – whose name we never catch – leads us out of the wood to a paved footpath overlooking a playing field, next to the school, with two conifers either side. We wait by the decaying trunk of a giant chestnut tree that collapsed after a storm a few years previously, crushing the (now rebuilt) wall behind us. As we idle against it, a constant procession of dog walkers stop to ask what we are waiting for. Everyone, it seems, has a story about seeing or hearing the new pair of ravens, even those who profess no other interest in birds.

Eventually our guide takes his leave. Vianet and I continue to loiter: he with his camera poised and me looking hopefully through my binoculars at the shape of each black jackdaw that flashes past; a far smaller and more common member of the corvid family, that is much more relaxed around people than its larger cousin. In London's Richmond Park, I have even fed them by hand.

That said, I've also done it with a raven on the top of Mount Seymour just outside of Vancouver, feeding two birds skulking about the viewpoint at the top of the trail with my niece's corn snacks. The ravens waddled over and stood so close to me I could admire the steel-coloured rings on their black talons and imagined I could hear the nictitating membrane flicker shut over their eyes. The Canadians on the mountain shrugged off my excitement, dismissing the birds as mischievous pests.

The Canadians have a far more relaxed relationship with the raven than their European counterparts. While the bird is steeped in First Nation culture, the raven is also regarded as a trickster and revered for its intelligence. Without being persecuted, and unburdened by allegories of omen and evil, the Canadian raven never needed to develop and pass on any real fear of humans. As I discovered, the bird will come as close as people allow it. There are stories I have read from medieval London, of ravens being fed by hand by children on the streets. As we grow used to the presence of each other once more, I wonder whether we may again revert to this distant age in our relationship with the bird.

After an hour or so of waiting, and numerous more conversations with curious dog walkers, it seems everybody in Nailsea has seen the ravens apart from Vianet and I. We wander back into the centre of the wood instead, admiring the saplings sprouting through cracks in the rocks and the brilliant emerald green hart's tongue fern fronds against the muddy banks. We clamber up to the top of an old quarry face, where some local youngsters have erected a rope swing over a terrifying 20 metre (65ft) drop, and dare each other to give it a go. And then I hear the sound that, over the course of researching and writing this book, my ears have grown ever-readily attuned to, a low, deep growl that heralds the presence of something altogether different. It seems to be coming from the conifer we have been unsuccessfully watching all morning, but a large blue steel fence means we can't get any closer. We rush back out of the wood to the footpath where we were previously stood and there we see the Nailsea pair.

One of the ravens slips out of the conifer first causing a rustle of branches to spring back under its weight. With a few heavy beats of its wings, the bird cuts over the field to land on a dead tree wrapped in eddies of ivy, 50m (164ft) or so away. The jackdaws beat a hasty retreat chittering among themselves in irritation as they do.

The raven stands on its perch – preening itself, looking about and snapping its great horned beak from side-to-side. It doesn't take any notice whatsoever of us standing there. Presumably the birds have grown well accustomed to their admirers in Nailsea. Unlike most other ravens I have seen in the wild (and apart from the first call I heard in the wood), it is completely silent.

After a few minutes or so the other bird – it is almost impossible to distinguish between male and female ravens from a distance – also takes flight and joins it on a nearby branch. Together they wait, surveying the options around them. Ravens are the earliest breeders among birds, and if this pair were to rear a brood this year, they would have already built a nest, although we cannot see any evidence of it in the treetops.

And then they take off, heading up with their diamond-tail feathers fanned out, banking around and soaring away from us in the direction of where the jackdaws scattered. They fly away in formation, so close they are almost wing-to-wing, casting a brilliant, black shadow against the clouds.

★ ★ ★

We hurry back to Bristol because we have another mission to complete before the day is out. As local author Sally Watson explored half a century ago in her book *Secret Underground Bristol*, the city is famous for its subterranean history. The Redcliffe Caves are a vast complex of man-made tunnels near the harbour, where between 1650 and 1750, huge quantities of red sandstone were dug out with shovel and pick, crushed down and mixed with minerals to make cheap bottle glass. Then there are the grounds of Goldney Hall in Clifton. Here, the eighteenth-century Quaker industrialist Thomas Goldney III built a grotto lined with the glittering carapaces of 160 species of shellfish – many traced back to the West Indies and Africa, and hint towards Bristol's darker past when it was one of the leading centres of British slavery. There is also the Clifton Rocks funicular railway dug inside the Avon Gorge, and used as an air-raid shelter during the Second World War, and the Hallen Fuel Depot, built to protect combustible tanks from Luftwaffe bombers. And then, there is the Ravenswell.

In 1370, the monks of Austin Friary began building a tunnel to divert water towards their friary in the Temple Gate area (where Temple Meads station stands today). The water they coveted was stored in a reservoir on the banks of the River Avon and fed by a spring, known as the Ravenswell. Over the centuries, this so-called 'temple pipe' was extended to accommodate the city's burgeoning population, before finally falling into disuse during the nineteenth century when it was surpassed by Victorian engineering. According to the local historian Kate Pollard, who wrote *Totterdown Rising*, Raven was a common surname throughout Bristol in the twelfth and thirteenth centuries, and the name of the well may have commemorated a local farmer. Across Bristol, there were three roads once named Top Raven Hill, Raven Hill and Raven Hill Mead. In the modern city, all three have been amalgamated into Ravenhill Road.

I had come across the Ravenswell during my research into the city's association with the bird and had discovered through various online caving forums that it was still possible to get inside. Some cavers had posted photographs from inside tunnels so narrow you can still see pick marks on the roof – caused during construction. I had gleaned that the spring was somewhere under the Totterdown Three Lamps, a cast iron signpost standing at the junction of the busy Bath and Wells roads. As for where the actual entrance was, all I had managed to get hold of was a crude map posted online by a man called 'Maggot'.

After Vianet and I get back from Nailsea, we stuff our waterproof trousers, head torches, video camera and wellies into the boot of my car and drive to Totterdown. We are not sure of the legality of what we are attempting to do and it is disconcerting when we arrive to see a police helicopter thundering above. As we park up on a side street, three officers run past us and another stops to ask if we have seen a man with 'long hair and no trousers' anywhere nearby. We apologise that we haven't, and nobody seems to bat an eyelid as we haul on our waterproofs and prepare for the caves.

My map leads us across the main road and over a wall, where a steep section of woodland drops down to the riverbank. We are close to the city centre, and the scrubby wood is covered in rubbish and fly-tipping; old plastic sheets and shopping trolleys fester, surrounded by dumped cider cans. It is satisfying to see springy saplings of juvenile alder growing all over the bank, in spite of the detritus. As we slip unsteadily down the hill in our wellies the drone of traffic behind us slowly quietens. Beneath us, the Avon churns.

According to the map, there are two entrances to the Ravenswell: one down near the riverbank and one by a railway bridge that spans the Avon. We head to the bridge first, but find any possible way in bolted up with thick, steel fencing, perhaps to deter the graffiti artists who have scrawled their tags all along the edge of the bridge.

Thwarted, we try to locate the riverbank entrance, which we presume is just below our feet out of sight. I suggest going down to the edge of the wood and then dropping down on to the mud and walking along until the entrance is visible. Vianet is hesitant because it's a drop of a good few metres down on to the riverbank with nothing then between the churning river and us. I insist all will be fine and decide to go first, with Vianet following behind.

I lower myself down from the trees as far as I can go and then drop down the last half metre or so on to the mud. As soon as I land though, my feet slide off and I fall on to my back. The momentum of my landing, and the fact I am wearing slippery waterproofs means as soon as I hit the ground I suddenly start sliding feet-first towards the Avon. I hear a woman across the riverbank scream, and all I can see is the brown water looming closer. I dig in the heels of my wellies and claw at the mud either side of me, in a desperate attempt to stop myself sliding in. Eventually, I come to a halt so close to the water's edge that all I can hear is the river rushing past.

The mud is a light caramel-brown and so thick that I have already been coated in it head to toe. I try and stand, but the soles of my wellies are too clogged up, so instead, I resort to dragging myself back up the riverbank to where Vianet is staring aghast from the edge of the woods above. From here I realise that I am going to have to climb back up to get to safety. I see in Vianet's face that this has suddenly become a serious problem, and he can see in my wide-eyed muddy stare that I have met with one of those singular moments in life where you realise the utter folly of your own blind optimism.

I slip and scrabble to where I can haul myself up, but even when Vianet reaches his arm down at full stretch, there is still a good metre or so distance between us. I grab on to the branch of a relatively strong looking alder to pull myself up and manage to get a foothold with one boot. But as I lift, the branch snaps and I fall back down into the river mud again.

Over the next 10 minutes or so I try several more routes up. With each failure, the ground gets more churned up and I become more exhausted. Eventually, with my arms weakening, I manage to get just high enough to grab hold of Vianet's hand and be hauled to safety. I clamber over him and wrap myself gratefully around a tree. As I lie there, mud-slicked and still gripped with fear, he does what every good filmmaker should; he reaches for his camera.

The mud, Vianet later tells me, is what eventually gets most of those who fall into the River Avon. You can fight the swirling currents for so long, but once the mud has you, there isn't much of a chance. I think of several passages I have read about ravens being so attuned to food bonanzas that they are often the first to discover drowned bodies. In 1870 a pair were seen pecking the eyes from the body of a drowned sailor under Boulby Cliff on the North Yorkshire Coast. In 1954 when the body of a male raven was dissected by biologists, a human lens, tendon and fingernail were discovered in its stomach. A pathologist confirmed the remains belonged to a drowned corpse that had washed up in Cumbria's River Eden. Had I slid just a fraction further, would one of Bristol's urban ravens have been the first to seek me out?

★ ★ ★

A while later I witnessed this nefarious hunting skill for myself. It was during a brief stay at the very south-western edge of the British Isles, on the Isles of Scilly. I had been sent by my newspaper to Tresco to investigate the unexplained disappearance of a young man who worked in a bar there and had vanished after a late-night party, only to wash up on a neighbouring island a few weeks later. I used to be a crime correspondent and, as a journalist, occupy a strange beat that can jump between birdwatching and suspected murder depending on the week.

As I loitered down near the island quay, trying to interview deeply suspicious locals on an island whose population exceeds no more than 175, I noticed two ravens circling over my head flying down towards a distant beach. When the day was over, I wandered over to where I'd spotted the birds.

On the fine, white sand, the body of a dolphin was washed up ashore. Its eyes had already been taken, leaving behind puckered sockets and around its tail and larynx were deep slashes right down to the bone. You could see where the ravens had been because of the long talon marks in the sand. The poor dolphin had also suffered more shallow slashes across its bluish grey skin, and here the sand was kicked up where gulls had scrapped between themselves. Wherever the ravens had stood to feed, however, there had obviously been no contest.

I later learn, through several press reports, that the dolphin is one of more than 60 to have washed up on the beaches of Devon and Cornwall during the first two months of the year. Conservationists have blamed the numbers on trawlers who fish in pairs with nets strung between them and then dump the suffocated bodies of the dolphins back overboard. Many of the washed-up bodies had been seemingly eaten by birds.

The *Sun* newspaper, in typical fashion, even started to question whether the 'mutilated bodies' had in fact been attacked by a shark. Any student of the raven knows the obvious culprit.

Even this far out to sea, the birds have a long association with the Scilly Isles. Indeed, to the north of Tresco is a wide expanse of beach known locally as Raven's Porth (Raven's Bay). Historically, there would have been one thing most vividly in common that the Scillonians and their ravens would have shared. The fortunes of these islands were part-built on the ships that came to grief on their outlying rocks. The Scilly 'wreckers', as they were known, were afforded a fearsome reputation among the maritime community.

Tresco's Abbey Gardens is home to a collection of shipwrecked figureheads collected over the centuries, while its teak and mahogany panelled interiors are also presumed to be made from wood salvaged from the wrecks. During the Scilly shipwreck of 1707, one of the greatest tragedies ever witnessed by the British Navy, when four ships and some 1,300 men were lost, a surviving sailor who somehow made it to shore, was rumoured to have been beaten to death by a local woman for his emerald ring.

Wherever humans tread, the raven follows. And while we do not like to admit it to ourselves, we have proved just as willing to profit from the death of another. Drowned eyes and drowned jewels glint the same to watchers from the shore.

CHAPTER FOUR

Speaking with Ravens

One November I wrote a column for The *Daily Telegraph* in which I made passing reference to my raven research. I described how a reader had recently made me aware of the author J. A. Baker's favourite winter walking routes, following peregrine falcons through the Essex countryside that he undertook in the process of writing his beautiful, unparalleled book, *The Peregrine*. I wrote that I too was attempting to beat my own path across Britain in pursuit of another bird.

Not, I hasten to add, that I dared in that column to compare my writing to Baker's starburst prose. I agree wholeheartedly with the former *Telegraph* editor and columnist Charles Moore, who describes *The Peregrine* as 'the most precise and poetic account of a bird – possibly of any non-human creature – ever written in English'.

Also, as Baker wrote with typical brevity, 'Time. Passes.' His book was intended as an elegy to the peregrine, which by the time of writing in the 1960s, had been decimated by the advent of pesticides in farming to fewer than 100 breeding pairs. The task I had set myself was, in that sense, an entirely different one. To examine our relationship – and particularly my own – with the raven. To unpick its ferocious intelligence. To ask why we find it so difficult within ourselves to regard the raven as a mere bird? To document and celebrate its return and to ask what that means in these troubled times.

There is another line of Baker's that sticks out in my mind from a passage where he has turned almost feral. 'The hunter has become the thing he hunts,' he writes. So the human and avian worlds begin to tangle.

The following month I received a message out of the blue. It was from a woman called Sarah-Jane Manarin whose mother had read my column. Sarah's email told me she had recently acquired a raven of her own: a two-year-old named Loki (after the mischievous Norse god), and so clever that he could play the xylophone and the children's game Kerplunk, as well as mimic human words. Would I be interested in meeting him? she asked. I replied in minutes.

A few weeks later, my wife and I catch an early morning train to Cheshunt on the north-eastern edge of London. Sarah and her husband Elliot run a falconry centre, Coda Falconry, on a farm a 45-minute walk from the station across the marshes of the Lee Valley.

I know this river well and used to come cycling along its banks on weekends as a child, long before the organic coffee shops and luxury flats started springing up. In that era, before the developers had moved in, there was space and scrub and barren, industrial buildings with rock-shot windows to explore.

Today, it is a cold morning, and a thick mist hangs over the water. Only the odd dog walkers are around, bidding us a muffled hello as they pass. Otherwise, the footpath is left to the robins and blue tits, fighting for their breeding grounds among the stark trees that have long shed their leaves. Starlings shoot like iron filings between electricity pylons that snake towards London through the fog. The marshland surrounding the river is still and steaming.

Sarah is there to greet us at the farm gates, beaming and ushering us in for a cup of tea. Avian flu is sweeping the countryside, so her birds are currently on lockdown. We dunk our feet in buckets of bleachy water before heading up the path towards her aviaries. Now in her early 30s, Sarah tells us that she was drawn to keeping birds following the death of her father nine years before.

When she was a girl, her dad often took her to birdwatching sites around Blakeney on the North Norfolk coast. As impressive as the flocks of waders were, she says, it was the birds of prey that got her hooked. She recalls the V-shaped soar of the marsh harriers and watching a kestrel hovering alone in the vast East Anglia sky, then diving down in a flash towards its prey. Later, she came to see a falconry display. 'It was from that moment that I wanted to know what made this trust between the falconer and the bird possible,' she says.

Sarah was 25 when her father passed away, and she returned to her home town of Harlow in Essex to try and better process the grief. 'I realised just how important it was to follow your dreams, as you don't get long in life to achieve them.'

That was the start of Coda Falconry. She decided not just to begin keeping birds, but also to do so in an urban environment where she could help educate children normally cut off from nature. She also takes some of her birds into local hospices to meet patients who can no longer get out into the wild themselves. 'Losing dad has made me stronger in the long run,' she says. 'I wanted to do him proud and be able to say I did my best.'

Falconry is a very traditional and male-dominated world, and Sarah says she has been castigated online for both being a woman and having the temerity to share its secrets with the underprivileged.

As part of her menagerie she keeps an eagle owl, a baby kestrel, four barn owls, a tawny and Sunda scops owl, six harris hawks, a peregrine, lanner falcon and saker falcon, an

Asian brown wood owl (which flies around her kitchen at home) and a magnificent snowy owl, whose burnt-orange eyes stare at us with a look of unblinking scorn as we arrive at the aviary.

At the furthermost cage lives Loki, the raven. A deep croak signifies that he has already heard us coming. We instinctively lower our voices and head into the office to hear Loki's story, before we meet him face-to-face.

He arrived, Elliot says, as a rescue bird. His previous owner was a man in Oxford, who was a lifelong keeper of birds but was too old to properly care for Loki, whom he had hand-reared as a fledgling raven. Loki ended up abandoned to a solitary existence in his outdoors aviary. He was eventually saved when the breeder who had sold his owner the raven, turned up, and was so shocked at Loki's condition that he took him away and put him up for offer as a rescue bird.

When Loki arrived at the centre the year before, his feathers were falling out and his left primaries had snapped off. His entire wedge-shaped tail – one of the prized indicators of the raven – had also disappeared. Once ensconced in his new surroundings, his instinct was to lash out, tearing with his beak at his buffalo-hide jesses and attacking anyone who came too close.

'He looked a state,' Elliot says. 'But emotionally he was a particular mess. He was incredibly angry and anxious and aggressive. He would be very violent when I came into his aviary, and there was a time early on in our relationship when he just went for my arm. His beak was stripping the flesh while he was hanging off of me. There was a lot of blood. I had never been attacked by a bird like that before.'

Elliot retreated, shocked at the damage, but the next day he returned. 'With a raven, it is all about hierarchy, and I thought if I gave him any sign that he had intimidated me then he would be above me in the pecking order. I went back in the next day, and he didn't go for me, but was very withdrawn. A couple of weeks after that, with me constantly being around, he realised he was below me in the hierarchy, and our relationship started developing.'

Despite their long experience training birds, the pair had never been near a raven before. Sarah says they soon realised Loki was a very different prospect. 'The eagles and hawks and falcons you can't really bond with on a personal level, but with Loki if he knows you're upset he will just come and sit with you for a while, rifle through your pockets and things like that. It is as though he can sense that things are wrong.'

When bird flu is not a threat, Loki often has free run of the farm. It did not take him long to work out how to unlock his aviary. Sarah says she normally gives Loki freedom to roam outside and he simply lets himself back in when weary of the world. The farm is open to the public and the previous summer she witnessed the moment a middle-aged man, walking with his son, came across Loki on the path. To her horror, the man aimed an unprovoked kick at the bird. Sarah raced down to confront him and says as soon as Loki saw her he jumped up into her arms for protection. The raven has remembered the attack and since croaked with alarm when other similar looking middle-aged men approach the farm.

'He really is just like a little person, and I stopped looking at him like he was just a bird a long time ago,' Elliot says. 'I would go as far as to say ravens are capable of love. The bonds they form are incredible. If you think love is about forming a trusting and loyal relationship, then that is what you get with ravens.'

A study released in 2017 by a research team from various European countries, further proved the ability of ravens to display human characteristics. Nine captive birds were given pieces of bread and then trained by the researchers to exchange those morsels for a piece of cheese instead – a far more

tempting snack for a raven. However, after establishing this
routine, some of the team then decided to occasionally cheat
the raven by taking the proffered bread and eating it, along
with the cheese, in front of the bird. The study discovered
that the ravens that had been duped subsequently refused to
trade with the people who had tricked them and instead
would swap with others not involved in the original exchange.
More still, the ravens remembered the faces of the researchers
who had tricked them for at least two months. This proved,
the research paper says, 'memory for direct reciprocity' in
ravens, and hints at one of the underlying impulses for their
complex social structure.

Aside from the ability to hold a grudge, I ask Elliot and
Sarah what other emotions they see in their raven? 'Empathy,
remorse, guilt, anger, fear, joy, anxiety, frustration … if he is
happy, then we definitely know about it.'

* * *

A while ago I had the pleasure of interviewing my childhood
hero, David Attenborough, for a magazine feature. We spent
an hour together discussing his own mortality (he was 90 at
the time), the solace that people find in the animal kingdom
and his fears at the environmental destruction he had
witnessed over his lifetime. We spoke about his latest series,
Planet Earth II, and one, now infamous, scene in the first
episode that seemed to encapsulate a lot of the themes we
were talking about.

It was filmed on the Galapagos Islands and showed baby
iguanas hatching and then instantly having to run for their
lives from racer snakes lying in wait. The scene showed the
snakes attacking individual lizards in their dozens, slithering
over one another to be the first to sink their fangs in. I told
Attenborough there was something of a gothic fairytale to

the footage, and that to me it seemed to represent a loss of innocence, with a life being snatched away so quickly. He agreed it appeared tragic but then added that we mustn't anthropomorphise these things.

And he was right, of course. The iguana hatching and snake attacking was a phenomenon that, no matter how shocking to the human mind, occurred purely out of instinct. There was no emotion at play whatsoever, and innocence had nothing to do with it. Purely a cold-blooded desire on both sides to survive. To seek to explain it in human terms was missing the point entirely.

This is a principle rooted in biological science, otherwise known as Lloyd Morgan's Canon after the great Victorian pioneer of animal psychology. The prerequisite being that: 'in no case is an animal activity to be interpreted as the outcome of the exercise of a higher psychical faculty if it can be fairly interpreted as the outcome of his exercise of one which stands lower in the psychological scale'. To put it simply: the minds of animals should never be compared to our own. As Attenborough insisted: 'we mustn't anthropomorphise'.

But when it comes to some species, mustn't we?

Arguably, this rule has held back studies of avian intelligence. In his foreword to Dr Nathan Emery's book *Bird Brain*, the Dutch primatologist Frans De Waal (who is the C.H. Chandler professor of psychology at Emory University in the US) says scientists have for so long carefully avoided the term 'cognition' in relation to birds, partly due to a fear that any contemplation of what was going on in the heads of animals was almost 'taboo'. New studies like those of Dr Emery, Senior Lecturer in Cognitive Biology at Queen Mary University of London, are beginning to find that birds (and corvids in particular) possess hugely more complex brain patterns than we first realised.

Over the past decade or so, several studies have shown corvids demonstrate cognitive abilities exceeding even the great apes. In 2009, Dr Emery presented a study where rooks appeared to be able to solve novel tasks through causal reasoning rather than simple trial-and-error learning – displaying a level of intelligence closer to a human baby than any other animal. Emery also mentions a separate study of 2013 that found ravens form their own political systems by remembering their long-term relationships, and at the same time intervening in others where they feel it benefits them.

Dr Emery is one half of a power-couple in the field of corvid intelligence. His wife, Nicky Clayton, is Professor of Comparative Cognition at Cambridge University. She started her career studying marsh tits but fell in love with corvids after researching the Eurasian jay. 'The thing I noticed when I was hand-raising baby birds was the marsh tits just begged at moving stimulus, such as the tweezers I used to feed them,' she tells me over the phone one afternoon. 'But I noticed the Eurasian jay always looked at people. They realised the moving stimulus is the tip of something much more significant: the tip connects with the hand, and the hand connects with the eye.'

A talented dancer, she is also scientist-in-residence at the Rambert Dance Company, and studies the intelligence of birds not just through their vocalisations but also their movements. She describes the way raven pairs fly together as a form of 'avian tango'.

'One of the things you see in closely-bonded pairs is their movement is so synchronous,' she explains. 'Everything in the bird world is sped up at least twice as fast, and their sense of space is phenomenal. When they fly closely together, there is literally a leader and a follower. We think it's a way of revealing and communicating that social connection.'

Her understanding of science and art means Professor Clayton is well aware of the tension over how to unravel the mysteries of the raven. 'Scientists try to objectify things,' she says. 'If you just objectify it you miss so much of what the dance or the behaviour or the cognition is about. There is a lot more to consciousness than just pattern and activation. Bringing in philosophy, art, history and culture allows you to have a much better-rounded picture and realise each one has its own strength and weaknesses.'

She says she gets contacted by a lot of people who tell her they have recently acquired a raven for a pet and find themselves stunned, not just at their intelligence, but also how they want to interact with humans. 'You look them in the eye, and you are so aware they are watching you,' she says. 'You realise that what lies behind that beady eye is a brainy, soulful creature. I can't say that scientifically, but that is my emotional response.'

* * *

We walk down the line of aviaries towards Loki's domain. Gimlet eyes pick us out in the gloom. The weird and wonderful parade of raptors that either stare or shriek as we pass reminds me a little of the scene in *The Silence of the Lambs* where Clarice Starling meets Hannibal Lecter in his cell. 'I've been expecting you, Clarice ...'

Of course, Loki has been expecting us. The office where we have been talking is well within the raven's earshot, and Elliot says he often hears Loki mimicking human words he has picked up from eavesdropping. Sometimes it can sound like a middle-aged man violently coughing at the end of the hall, or a high-pitched laugh. During last year's lambing season Loki even learnt to mimic the sound of a ewe giving birth. It reminds me of a tale I once read of a raven at Chatham Barracks in Kent, who sent the soldiers marching out on parade because he had mastered the cries of the sergeant major in charge.

Elliot, who is a filmmaker when he is not working with the birds, is wearing a flying jacket filled with pieces of cheddar and crumbled chocolate Hobnobs. As he walks into Loki's aviary, he produces a bit of cheese, which the raven snatches away and stores in the bottom of his beak without swallowing it. Next, he produces a peanut tin with another morsel inside. Within a matter of seconds, Loki has managed to prise off the lid and gobble it down. Again, he keeps the food in his beak without swallowing, intending to stash the food away somewhere when we are not looking. A long line of drool drops from the raven's beak on to the floor. I have never noticed this in the birds before, but the saliva they emit is a hallmark. There is also a smell familiar to me from when I have previously seen the birds up close: the musty, slightly cloying odours of a raven's eyrie.

His other raptors, Elliot says as we watch Loki hop about his cage, are only motivated by food. The origins of the phrase 'fed up' lie in the falconry world; when a bird that has eaten its fill ignores all instruction and refuses to fly. Loki, by contrast, requires constant stimulation. When he is bored, he will terrorise the birds in the cages nearby; teasing them with food that he dangles just out of reach. Captive ravens also do this with domestic pets, tormenting them by tweaking their tails and flying off. It is believed this teasing is all done for the same reason a human would: to establish their own place in the hierarchy.

He is startlingly tactile with Elliot, jumping up on to his shoulder and head and allowing him to stroke his black, glossy plumage. It takes a few minutes for him to settle and grow used to the strange faces looking in at him, then he permits us to stroke his soft neck feathers too. This close, the raven's plumage is an array of purple, greens and brown that pool shimmering together. His ear feathers stick up, a sign he is enjoying the contact, and his throat and head puffs out.

Once Loki is relaxed, Elliot carries him out on his arm to weigh him on the scales as he does every day. Today the raven is 1.3kg (2lb 8oz), slightly over his normal weight. As we walk towards an outbuilding where Loki's xylophone and Kerplunk game has been set up, the raven cries what previous studies of raven linguistics have deemed the 'gro-call', used by the bird to establish itself in social contexts. Elliot describes it as a 'bah' instead. Either way, as we leave, it sets off the baby kestrel into a fit of shrieking. No doubt Loki the mischief-maker would be pleased.

We are told the raven has never been in the room where we head, a rectangular-shaped classroom for the children with chairs in the middle, metal lockers at one end and grass and flowers painted on the walls. Inside are Sarah and 19-year-old Emily, who works at the falconry centre.

The first thing Loki does when he arrives in the room is to jump off Elliot's hand and fly up on to my shoulder. He makes the movement in less than a second and before we all know it he is already up there. Presumably, he picks me because I am the tallest in the room, so it gives him the best perch to scan his surroundings. The speed with which he moves reminds me of what Nicky Clayton told me about the speed of the raven brain – that the world they occupy moves twice as fast as our own.

The raven is surprisingly light on my shoulder. Loki's talons tighten as he balances, but do not dig in to my skin. When he has assessed the room and those in it, he hops down on to the floor and begins waddling about with his bloomers splayed. He recognises the xylophone (a child's plastic version with just five keys in bright, primary colours), jumps up on to the table and starts hammering away at them with his beak. Elliot offers a ruby-red chicken tendon from his flying jacket as a reward.

Next, it is time for Kerplunk, an empty, plastic bottle the size of a litre with skewers speared through it, one of which holds a dead chick. Sarah warns me that it has been several months since Loki last played the game so he may have forgotten what it is, but as soon as the bottle is produced the raven sets about it. One by one he pulls the sticks out and drops them on the floor until the chick is released. He scoops it out of the bottle and hacks it apart with his giant beak. The only piece of the animal that he leaves is its stomach, which he removes with surgical precision. The rest of the chick is gulped down. I have been recording Loki in action and check the time on my phone: he had done the whole thing in 39 seconds.

It is fascinating watching Loki explore the new room. When he thinks we are not looking he pokes his head inside the metal lockers to see what he can find, he also notices the grass and flowers painted on the walls and pecks with his beak to check whether they are suitable caching spots to hide food.

Once he has established his surroundings, he marches towards me, cocking his head so he stares directly into my eyes. It is oddly difficult to maintain contact with the raven's stare, over which its nictitating membrane snaps back and forth as he contemplates us in turn.

The eyes mean everything to the raven. As Derek Goodwin notes in *Crows of the World*, when hostile, a raven's eyes become wide, and pupils dilated. When friendly, as Loki seems now, the eyelids are slightly closed, and pupils contracted. Despite the eyes being the first thing a raven pecks out when it falls upon prey, when the birds fight among themselves they never attempt to blind each other. The raven understands the significance of what the eye can do and also what lies behind it. According to Goodwin, captive birds also respond appropriately to friendly, aggressive or intimidating looks by humans.

Because Loki was hand-reared, Elliot says, he regards himself as human and as a result, has established an appropriate hierarchy around the farm. All of the other birds and animals are below him, and so he takes the opportunity whenever he can to tease them and show his dominance, in the manner of a schoolyard bully. He has placed himself on a par with Emily and regards Sarah and Elliot at the top.

Intriguingly, Sarah tells me, she recently noticed wild ravens flying over the farm, although Loki did not apparently hear their kronks as they passed by. I am excited by the fact the birds are coming so near to London and resolve to try and establish just how close the raven is to returning to the heart of the capital.

It is clear Loki can also differentiate between human genders. His call when he is responding to Elliot is a much deeper rasping sound, while when he addresses Sarah, it is far more plaintive. At one point, he dips his head and stretches his wings down to the floor in a long, elaborate bow in front of her. I later discover this is a typical movement a raven pulls off when it wishes to show submission.

When it comes to Emily, though, Loki is a terror. When he grows bored in the room or his attention drifts, he hops over to her and starts pecking at her hand and pulling at her clothes. She tells me while trying to bat him away, that he sees theirs as a sibling rivalry trying to vie for the affections of the parents.

I am writing all this down in my notebook and wonder aloud to the room, whether Loki could take the lid off my biro. At hearing his name and seeing me raise the pen the raven spins around and flies up to my knee. He pecks several times and when he gets frustrated that he cannot get it off, nips me gently on the knuckle. I loosen the lid a bit and give

him another go. Within a few seconds, he has managed to do it. Then, after inspecting the object, he scuttles off again.

The most interesting of our experiments with Loki comes a week or so after our visit. Elliot and I have discussed setting up recording equipment outside of Loki's aviary to capture some of the strange sounds Elliot hears him mimic when he and Sarah are sitting up late. Elliot emails to tell me he has set up a microphone for two successive nights and will send over the results.

When the recording arrives, Elliot has managed to capture the most curious, unexpected thing: silence. Only when Loki knows he has an audience within earshot does he bother speaking up. Otherwise, he retreats into his own impregnable thoughts. We mustn't anthropomorphise, but how then to explain a raven that can shift so readily between the human and avian worlds?

CHAPTER FIVE

Ravens and the Forest

Ravens take you to other places. I am walking in the fading light along a railway line, as trains running from Ashurst to Brockenhurst thunder by at 20-minute intervals, first north, then south; bearing passengers down towards the Solent. Their headlights light up the gloom and horn blasts shake me as they pass one another at gaps 45 seconds apart. I wonder if anybody on board sees the lone figure in a blue jacket trudging by the towering Douglas firs. It is late February and the days are closing like drunken eyelids.

I am heading towards an X marked in pencil on my Ordnance Survey map of the New Forest – a distant part of the national park called the Ladycross Inclosure. This is a different forest from the one I have been tramping through in recent days. There is no footpath for a start, only a vague beaten track flanked by barbed wire and through gorse bushes (known locally as furze) and sucking bogs. I haven't seen another soul walking for miles.

I leave the railway line at a signal box fizzing with electricity, then, as the light switches from green to red, disappear deeper into the silent wood. Unlike the carefully nurtured ancient copses elsewhere in the forest, this is a commercial operation, with deep scars in the stretches of conifer where they have been felled for timber. At points, it appears as though a convoy of bulldozers has bored through the trees.

Twice I stumble into a bog, and twice I yelp as binoculars rattle my ribcage and the water sluices in over the tops of my boots. Thick, dirty-grey clouds swirling among the canopy of trees swallow the noise. I squelch on until I locate the stream on my map that will take me to my intended destination. The treescape slowly changes into gnarled old birch growing thick together and neglected by the loggers. This is the sort of place where one contemplates the tangle of roots far below the earth and the subterranean cathedrals upon which we stand.

There are two crows calling to one another from trees spaced 50m (164ft) or so apart. I walk through the middle of their conversation, hearing their voices rise and feeling eyes regarding me from either side. Night is now too near for me to see what I am after – the long surfboard of twigs hidden at the top of the tallest trees that indicates a pair of nesting ravens. So, instead, I sit on the trunk of an upturned, mouldering beech, teeming with platoons of insects hell-bent on finishing their work for the day, and I listen to the encroaching darkness.

Because of their scarcity in recent memory, we presume ravens to be birds of the extremities, building their nests on cliff faces and pylons out of sight and largely impossible for us to reach. Left to their own devices, however, they are equally happy as birds of the forest. Many of the first great naturalists of the eighteenth and nineteenth centuries – the Reverend Gilbert White, Charles Waterton and John Clare among them – wrote of raven nests built high up in the most inaccessible wooded citadels and constantly targeted by villagers as a result; either for the honour of being the first to reach the nest or simply to force the ravens away. In his poem *The Raven Nest*, the poet John Clare wrote:

> Not one is bold enough to dare the way
> up the monstrous old oak where every spring
> finds the two ancient birds at their old task.

The raven was one of the many original occupants of the great forests that covered Britain 10,000 years ago at the end of the last Ice Age. The first trees to begin replacing the tundra left behind on the retreating ice shelves were the pioneer species; those seeds borne far and wide on gusts of wind. Hardy birch is thought to have formed some of the first forests, followed by aspen, hazel, pine, elm and oak. Our culture and sense of self are shaped around these trees. There is a deeply ingrained sense of romance *and* fear of the thick, impenetrable woodland stretching from coast to coast that permeates so many of our stories. The forests are where our heroes are forged, but also where nightmares lurk.

Some 40 years ago, the great English author and naturalist Oliver Rackham coined the term 'the wildwood'. TH White called it 'the Forest Sauvage'. In White's Arthurian epic *The Once and Future King*, the young King Arthur wanders through this enchanted world followed by raven eyes as well as wild boars and wolves. For White, all three animals represented the wild side of our national character and that of his own. When Archimedes the owl lectures the young Arthur on the various qualities of the birds around them, he pays particular attention to the raven: 'the oldest, gayest and most beautiful of all the conscious aeronauts'. White was a keen ornithologist; a trainer of goshawks and obsessive over birds whose currency is blood. 'They talk every night, deep into the darkness,' Merlin tells his pupil, of the raptors kept by the falconer in the castle. At dusk in raven woods, those same ancient conversations are played out. Guttural croaks that sound as if from another time.

There are many other more ancient fables linking King Arthur to the raven. One long-established folk tale has it that following his mortal death, the soul of King Arthur escaped in the form of a raven and one day will return to save the kingdom from its avaricious rulers. Numerous books on English superstition, including Robert Hunt's 1865 *Popular Romances of the West of England*, repeat the well-worn story of Marazion Green in Cornwall, where a hunter who took casual aim at a passing raven was scolded by an old man for attempting to shoot a bird possessed by the Arthurian spirit.

The Ancient Britons revered their wild woods, attributing to the life inside them deep religious significance. They regarded trees as symbols of death and regeneration and imbued each with a particular spirit. Over the past thousand years we have reshaped this island in our own image of progress, and the Wildwood, along with the creatures that lurked there, has been hacked back into the past. Despite the medieval legend that it was possible for a squirrel to travel across Britain from the River Severn to the Wash without touching the ground, by then the decline was already terminal. The raven trees were axed down and their roosts riddled with bullets. Ornithologists pilfered their eggs in the name of science, and opportunistic collectors in the name of commerce.

Today so-called ancient woodland (defined in England as in continuous existence since 1600 AD) now only covers around two per cent of Great Britain. The New Forest, which spans nearly 38,000 hectares, is said to best resemble our now long-lost woodlands in both species and scale. Odd then that the forest itself was an imposition on the natural environment, planted some 1,000 years ago entirely by human hands. It was in 1079, 13 years after victory in the Battle of Hastings, when William the Conqueror sequestered this swathe of Hampshire as a king's pleasure ground. Primarily, it was so he and his court could hunt and slaughter 'beasts of the chase' (deer and wild boar) across what had been open grassland and bog. While there is evidence of cultivation dating as far back as the Bronze Age, the New Forest is primarily an infertile area. The Normans set about planting great forests of beech and oak (which in later centuries were felled to build ships for Nelson), and strict laws were imposed protecting hunting rights above all. Those commoners who previously lived on the land and now defied the Crown were brutally oppressed.

Over the ensuing centuries, the forest, its wildlife and people were pillaged by successive nobles until the first inclosure acts began in the late seventeenth century. These were designed to protect parts of the landscape for viable timber production but led to the formation of the New Forest as it is today. Those living and working here still insist on this traditional spelling of 'inclosure'. As with so much of the New Forest, its language remains deeply rooted in ancient terms. A pool of open ground is known as a 'shade', and a clump of trees a 'hat'. Beechmast and acorns are referred to as 'the turn-out', stumps are 'stools', roots are 'mocks' or 'mootes', and the wind whistling through the trees is known as the 'hooi'.

These old trees once thronged with ravens, drawn perhaps in part by the daily bloodshed and easy meat from the hunts. One of the earliest inclosures established under the first New Forest Act of 1698 was called Raven's Nest. By the nineteenth century, however, the ravens had been wiped out altogether. There are two conflicting dates for the last recorded nesting of ravens in the New Forest. The Victorian naturalist and writer, John Richard de Capel Wise, puts it at 1858 when the last two nests were taken in the old woods around Burley by local youths armed with sticks and stones to drive the furious raven parents away. The Hampshire Bird Atlas tells it slightly differently, relying instead upon Kelsall & Munn's 1905 book *The Birds of Hampshire and the Isle of Wight*, which puts the last recorded raven nesting at 1887. Persecution, exacerbated by the switch from sheep husbandry to cereal cultivation, was blamed on the decline. Between 1951 and 1992 there were just 33 birds recorded. These were nearly all from around the coast and thought to be individual birds making forays across the Solent from the Isle of Wight where there has remained a stubborn breeding population. This short hop to the mainland aside, ravens are known for their dislike of long flights over large bodies of water – although rare individuals have been recorded flying up to 500km (310 miles). This desire to stay over land is perhaps why so many countries – and especially this island nation – have such a close and complex relationship with the raven. Despite the huge geographical spread of the species across almost every continent, the birds remain unique to each land mass they reside in, and instinctively learn and mirror the behaviour of its populations.

Since the 1990s the frequency of raven sightings in the New Forest has increased exponentially: 69 between 1995 and 1998; 91 in the year 2000; 258 in 2002 and in 2004 a pair fledged four young from a pylon nest near Fordingbridge on the western edge of the forest. In 2008, 14 nests were found across Hampshire, and pairs of family parties were noted at a further 23 locations. In 2014, reports of ravens came from 360 separate 1km squares mapped in Hampshire compared to 230 the previous year. There are presently three known nesting sites from within the heart of the New Forest that I have been told about and marked on my map.

To find them, I have booked a dormitory bed in a hostel in the middle of the forest, which at this time of year is shared with contractors working in and around the area. At night we talk little, each exhausted by our own particular tasks and every grey morning the alarms on our mobile phones chime at 6 am almost in unison. Then I fill my Thermos with tea and strike out into the forest alone to try and find the raven nests.

The Ladycross Inclosure contains one of the most recent nesting locations, but as I sit hoping for the ravens to reveal their home somewhere above me in the trees, I begin to realise it might be a futile end to a long walk. At one point, after 30 minutes or so of waiting, a pair of crows take off in jabbering flight. Perhaps this may mean their larger corvid cousins are re-entering their territory. Instead, it turns out to be the magnificent silhouette of a sparrowhawk ghosting from tree to tree on an evening hunt. I watch its wings flap twice and glide; then the hawk is swallowed up by the forest. Night settles around us, and after a time I too take my leave.

Walking back disconsolately towards the train track and my single dorm bed, I see the signal box has switched to red, and hear the northbound service once more thundering up the line. Clattering noise, then 20 minutes of stillness. The modern rhythms of the forest clashing with the old.

* * *

I have been informed of the tree nesting sites by a man called Andy Page, a former gamekeeper and now the Forestry Commission's Head of Wildlife Management for the New Forest, who has discovered them himself on his daily rounds. I meet him the day after my failed attempts to locate a nest in the Ladycross Inclosure.

Aged in his late 50s and with 20 years' experience working in the New Forest, Andy Page is a rarity in conservation circles: somebody with experience of 'the other side'. Normally gamekeepers, and those monitoring wildlife, are accustomed to crossing swords with each other over countryside management issues and particularly when it comes to birds of prey. The need for 24-hour monitoring of the final few hen harriers in the Forest of Bowland for fear they may be blasted out of the sky is no greater indication of this sometimes unbridgeable gulf. Page, weathered and healthy from a lifetime spent outdoors, and who still speaks with a gamekeeper's gruffness, first began working on a nearby country estate in the 1970s, motivated, he says, by a fascination with the 'relationship between hunters and their prey'. 'When I started it was still the bad era of anything with a hooked beak was dangerous and needed to be killed,' he tells me. 'I never agreed with that, but when you are young, it's difficult to influence things. I saw illegal persecution of protected species. I really didn't like it so came out to study raptors instead.'

During his time working on the shooting estates he never saw ravens, the guns had put paid to that. It was the same when he moved to the New Forest 20 years ago. As time passed though, Andy started receiving reports of the occasional sighting: the birds were coming back. Each occasion was so rare that he and his colleagues would make a note of it. Then, a few years ago, he found two raven nesting sites in the forest. The following year he stumbled across a third.

We meet in the village of Lyndhurst, one of the centres of the New Forest and home to its most vivid juxtapositions. On the high street is a Ferrari dealership with gleaming red engines on its forecourt to attract the increasing number of second-home owners moving in, while nearby is the old Court of the Verderers, which dates back to the thirteenth century and still sits on the third Wednesday of each month. Nowadays, it mainly deals with managing the donkeys and ponies that graze on the common land outside the forest inclosures.

We rattle off in Andy Page's green 4x4 heading for a few potential sites where he has previously recorded pairs of ravens. Attached to the sun visor, he keeps the moulted feathers of various birds gathered from the forest floor:

sparrowhawks, a goshawk plume shaded like the crests of rising waves, though not yet a raven, this year at least.

The birds, who mate for life, are famously early breeders. From February pairs will most likely be on the nest – and this is when they are at their most secretive.

On the way, Page suggests a detour to a part of the forest called Amberslade Bottom where, that morning, he and a few other Forestry Commission marksmen had been stalking fallow deer as part of the New Forest's annual cull. I was also up early in a different part of the forest that morning, wandering through an old copse of oaks overlooking a stretch of moorland known as Backley Plain in the hope of seeing ravens flying over from their morning roosts. Instead, I found myself entranced by a deafening dawn chorus led by chaffinch and green and goldfinch. I watched them flitting between the oaks, as the sun broke through on to a hard frost that had settled overnight; reddening the earth and making steam plume from the forest floor. 'Haze-fire' is the old poetic description for it: the bright, blinding light of a dawn sun and luminous mist rising. With the warmth on my back, it caused a momentary state of joyous paralysis, and Page tells me it is the same with the deer. On fine days, they drift more slowly to their woodland couches, pausing just a few moments longer to soak up the rays catching on their ginger and starlit fur. 'At this time of year, they start an hour earlier and go to bed an hour later,' he says. 'The odds are tipped slightly in our favour.'

That morning, seven deer made the fatal mistake of momentarily dropping their guard, oblivious to Page and his colleagues crouched among the still-frozen bracken, each waiting for the sight to steady on their .308-calibre rifles and a clear shot to the heart. We arrive at the scene of the shootings and walk down a steep bank into the forests where the deer corpses have been gralloched with a hunting knife. In skilled hands, the process takes no more than 30 seconds; the bodies cut open, and stomachs pulled out with a twist of the oesophagus. Each body is dragged away to be butchered for meat, while the stomachs are discarded, left cooling and swelling on the bracken for scavenging birds to eat. In their white casings, they look like Beef Wellingtons.

Each year in the New Forest some 1,000 fallow, roe, sika and red deer meet this fate in the annual cull. Once, their ritual slaughter was in the name of sport, but now it is in the name of conservation. The over-sized deer herds cause extensive damage to young trees, stripping the leaves in spring and munching off layers of bark in winter forage. In summer, young males split saplings; using their fragile branches as rubbing posts to scrape off the velvet fuzz on their antlers and harden themselves up for the rutting season ahead. Ever the opportunists, where once the crack of a rifle would send ravens scattering, now in the New Forest they know the sound indicates an easy meal. Page tells me the birds sometimes start to follow the hunt before it has finished and, if a shot does not fell a deer instantly, ravens can reach the deer's carcass before they do. 'It happens when we have been stalking the deer for a long time,' he says. 'That is when the ravens get them. They know to go straight away for the best meat in the haunches, and if they do, they spoil the carcass, for human consumption, at least.'

We follow a line of blood through the undergrowth that leads to the stomach of an eight-month-old fawn loosely covered by bracken. It has so far been undisturbed by ravens, but we decide to wait nearby overlooking the spot where Page also spotted a great grey shrike that morning. The bird is a rare and solitary winter visitor, and even in 20 years, he has only seen it here a handful of times. For a while, we

quietly kick our heels in the bracken, but both the shrike and ravens seem to have business elsewhere. Eventually, we decide to take our leave. As we emerge from the forest, other shapes loom on the banks above us. A handful of twitchers, who have already heard of reported sightings of the great grey shrike, has arrived from the coast and are setting up their telescopes poised for it to reappear. We stop to talk to them and they tell us that several ravens have been spotted flying in our direction from a stretch of bleak moorland known as Black Gutter Bottom. Perhaps the twitchers' extended telescopes caused the raiding party to veer away at the last minute. For birds as intelligent as the raven, even the scent of danger trumps that of blood.

We drive between a few other potential nesting sites. Both are similar in layout to where the ravens have settled in Ladycross Inclosure. The huge, metre-long nests are built high at the top of trees, yet not quite at the top to try and afford as much protection from the wind as possible when 20m (65ft) up in the air. The birds also prefer to nest on the forest edge to identify potential threats long before they arrive. As we scan the Douglas fir for signs of life, Andy Page ruminates gloomily that the nests may well have been dislodged over a particularly stormy winter. I think back to his description of the early part of his career when ravens were non-existent altogether. Recovery, like the nests they build, is an impressive yet fragile thing.

★ ★ ★

John Richard de Capel Wise lived here in the 1860s when the old ways in the New Forest were most under threat from modernity's march. He would take long walks with the young illustrator Walter Crane documenting the scenery and wildlife they encountered. I decide to adopt the nineteenth century explorer's approach to try and finally locate some resident ravens within the New Forest. He not only urges the visitor to wander through the forest on foot, but also 'to follow the course of one of its streams'.

The streams and springs here are part of New Forest folklore. There are around a dozen recorded major river systems spilling into a vast network of streams with beds of gravel, sand and clay. Iron's Well, a chalybeate waterway near Fritham that gets its rusty red colour from the iron deposited along its banks, was known as the Leper's Well for its supposed medicinal properties. Other ferruginous springs in the southern part of the forest are similarly renowned.

I start on a sodden morning from the old oaks of Burley Wood. Woodpeckers drum their dawn beat into the fissures of the trees as I take my first squelchy steps along a small gravel-bed stream, gin-clear and littered with stones the size of a fingernail and pale as ivory. A few oblivious donkeys drink from puddles as I pass by. A chorus of chaffinch, goldfinch and wrens welcomes me towards the wood, but then something else takes over. From inside the trees I hear the unmistakable kronk of a raven. A scattering of crows that have been perching precariously at the top of a clutch of silver birch, take off in pursuit. That deep sound echoes thrillingly in my ears before the normal dawn murmurings of the forest resume. I never manage to see the raven, and further inside, raptors take over: another sparrowhawk and then later a buzzard gliding silently over a flooded marsh before settling in the upper boughs of a pine.

A few miles away on the map is Norley Wood, an inclosure close to the marshes of the Boldre Foreshore and the eighteenth-century shipbuilding village of Buckler's Hard. In 1811 they decided to create an inclosure protecting 163 acres of Norley Wood's remaining oak, elm and beech forests. I make for its southern point to a place called Ravensbeck Farm. The farm lies at the end of a road past well-kept gardens of expensive-looking cottages, and bird tables teeming with sparrows whose chirps intersperse with a woman singing opera in her living room. Everything, from front doors to border hedges, appears freshly-painted and clipped, although Ravensbeck Farm at the end offers a different prospect. The farmhouse is brick-built with its walls coloured a pale green wash. Up a muddy path there are large, corrugated cowsheds on one side, and stone troughs on the other sprouting early season English bluebells and daffodils. As I near the front door, I am assailed by a medley of cocker spaniels and sheepdogs yapping and growling.

A man wearing a fleece and a curious look emerges; he is Clive House, and his family go back generations on the farm. The Houses are, he tells me, the last tenant farmers in the area. He took the business over from his father Bill, and his 25-year-old son Tom works nearby at Milford on Sea, where he keeps sheep and lambs. At Ravensbeck, Clive House keeps cows. When I tell him of my interest in ravens and ask about the provenance of his farm's name, his affable features cloud. Ravens reappeared about 10 or 12 years ago, he tells me, after never previously straying anywhere near the farm, at least in his lifetime. 'I first noticed them on the farm where my son works,' he says. 'This seagull was on the field with its wing broken and then these ravens came down and just mobbed it. They stove its head in something terrible. I just thought to myself, that isn't fair. He has sheep there and lambs as well. The ravens land on them and take their eyes out. When there are three or four of them on a little lamb it doesn't stand a chance.'

House says he has also experienced problems on his own farm, with ravens worrying his cows when they are giving birth; attempting to eat the placenta while it is still attached. Recently, he has started calving inside the corrugated sheds to better protect his herds. 'They can smell the blood,' he says. 'They always fly off before you can get too close. If there are only a handful then it is fine, but too many and it can start becoming a problem. They do look quite evil when you see them up close and when you hear them as well. The feeling is we have got enough at the moment. We don't want to see numbers increase by much more because then we have got a problem.'

His words echo in my head as I bid farewell and head up into Norley Wood towards another raven's nest. Tales of raven savagery are common enough among farmers, but still, the details are disturbing. There is something in the way the birds watch and wait for an opportunity. The fact that when they strike, they know they only have a few moments and go for the softest, most vulnerable parts – the eyes and haunches to maximize the kill before humans can intervene. As with the birds stalking wounded deer elsewhere in the forest, they have adjusted to our behaviour in order to feed and thrive. It is an uneasy thought to dwell upon, as I strike off alone into raven woods.

I approach Norley via a stream that leads up into Wormstall Wood and has burst its banks in recent days, leaving me hopping between patches of firmer ground and sinking several times into the morass. Soon, after taking my leave from Ravensbeck Farm, a hail shower breaks out, littering the ground with beads of ice. The going is difficult, so I take with me a stick I have found nearby from one of the alder trees that line the stream's bank.

During his wanderings through the New Forest, Wise writes of numerous encounters with gypsy families living among the trees, their 'brown tents always fluttering in the wind' and their nightly camp fires reddening the firs as they baked squirrels and hedgehogs in clay pots. A study of Romani gypsies conducted in 2012, discovered that the roots of Continental Europe's largest minority group stretched back 1,500 years ago to their arrival from India. In Britain, the first travellers are thought to have arrived somewhere around 1513. The wild woods of the New Forest appealed to families used to both persecution and living off the land, and Romani families soon established themselves in Hampshire. New Forest gypsies were known as 'Nevi Wesh' and their homes, known as 'benders', were made out of frames of flexible hazel wood woven into a dome shape and then covered with blankets and sack cloth. To make a living, they poached game and wove mats of briar and heather, as well as baskets, brooms and straw ropes.

As with the ravens, a period of 'civilisation' of the forest beginning in the nineteenth century slowly drove the Nevi Wesh away. Wise suggests the Deer Removal Act of 1851 (where a concerted, if ultimately unsuccessful attempt, was made to flush all deer out of the forest) cut off their main food supply. At the same time, various Christian missionaries tried to persuade gypsy families to leave the forest, while officialdom adopted a somewhat more heavy-handed approach. In 1913, Lord Arthur Cecil complained to the House of Commons that gypsies were straying within 100 yards of the manicured gardens of his sixteenth-century country pile Passford House, not far from Lymington. A decade later it was decreed that all gypsies in the New Forest were to be moved into seven encampments, no longer permitting them to roam where they wished. Following the end of the Second World War this was reduced to five.

There is a British Pathé newsreel from 1948 documenting the squalid conditions gypsies were forced to live in within the New Forest compounds, described in the film as 'primitive as the Stone Age'. 'Deep in the New Forest in a fairytale setting where beauty and squalor sit side by side, a new kind of squatter has set a problem for the authorities.' So begins the narrator, in his clipped broadcasting English, over a stirring soundtrack of strings; whilst the camera pans across the tents which 18 families call home, and women unpeg washing from lines strung between the trees. 'The gypsies say,' the narrator continues, 'where can we go, but the woods?' By the 1960s that decision was made on their behalf, and the last Romani families were moved out of the woods on to council estates.

On my way into Norley Inclosure I pass Broom Hill where one of the seven gypsy compounds was built. Nowadays, it is covered by holly shrubs and populated by the odd mistle thrush noisily guarding its stash of winter berries. There are well-trodden footpaths for dog walkers and family hikes, and a steady pattern of horseshoes is squashed into the mud. In the distance, across a swathe of Beaulieu Heath, I hear traffic trundling along the B3054. I think of the newsreel narrator's throwaway claims at what we regard to be beauty and squalor, the ghosts who passed their lives among these trees, and the wild ways we have taken it upon ourselves to stamp out.

Eventually, deep inside Norley Inclosure, I locate one of the raven nests in a thick copse of old pines. It is roughly a metre long and built across two branches close to the top of the tree some 12m (40ft) above. The nest is expertly woven, like the gypsy tents, and only visible from a certain angle. I stake it out but there is no sign of a resident pair. Still, I am happy enough listening to the yaffling green woodpeckers nearby, and finding myself in a quiet corner where the old wood remains uncivilised.

* * *

My final morning of a week in the forest and I am sitting by a farm gate looking out over a field below containing a hundred or so pig pens. At the top end, roughly 500m (1,640ft) away from me, two men hammer in fence posts while their tractor stereo plays David Bowie's *Oh! You Pretty Things*. There are a hundred or so domed pens scattered about the field that, from a distance, resemble corrugated air-raid shelters. In each lives a single well-fed sow and her piglets, and on the top of at least half of the huts, is a raven.

Often, they are in pairs, with another bird lurking next to the entrance just out of sight of the pigs. Around them swirl flocks of black-backed gulls and crows, but it is the ravens that dominate the field – perched on post and pen and any other position of authority – while the crows contest the ground. From my well-hidden vantage point, juvenile ravens keep flying past me, croaking like the pigs they covet, and jarring with the distant strains of the Bowie song. Down on the field a few tip their wings in aerial display. After days searching in vain for solitary birds or pairs I am astonished to see so many.

I came to the New Forest with the intention of finding ravens nesting in trees, as they had done for millennia, but instead I have found them thriving in a far more modern setting. After meeting Clive House on Ravensbeck Farm and hearing how the burgeoning bird populations had started to trouble farmers, I decided to ring around a few more of those who worked on the pastureland over Tidpit and Martin Down, a chalk mound overlooking the forest's eastern boundaries. On one single day in May in 2012, 105 ravens were counted by a volunteer helping out with the Hampshire Bird Atlas, and it is a figure I am keen to check. Eventually, I am put in contact with a man called Rob Shepherd. When I tell him the reason for my call, there is a pause at the end of the line. 'That must have been a quiet day,' he eventually says.

The next morning, we sit sipping coffee in his farm office while rooks croak from the trees outside. Even driving towards the farm, I have spotted ravens perched along the road and dotting his 2,000 acre fields like black blots of ink. His father, Roy, started up on this land in 1963 as a tenant to the eighth Earl of Radnor. Both Roy and the Earl have now died, and their respective successors have continued the partnership. Shepherd grew up on the farm and has witnessed huge changes to the pressure placed upon land that, until recent history, was regarded as relatively infertile, yet in the modern era, is relied upon to constantly produce. During the Napoleonic Wars, he says, these hills turned from sheep grazing pastures to producing barley to feed Britain's armies. The world wars of the twentieth century meant even more of the land was ploughed up.

'When I was young only spring crops were planted in March as the main crops grown on this land,' he says. 'In the 1970s, it became normal to grow winter crops, which were planted in autumn. This meant the land had very little time when it didn't have cereal crops growing. It was also the time of the mechanization revolution, where we were trying to produce as much food as we possibly could and were going full speed. This meant basically that nature and conservation very much took second place. We were given money and grants to take hedges out, and all sorts of grants to farm in ways not conducive to wildlife. That came from the government. Between 1945 and 1946 most people in Europe were eating grass they were so hungry. Food became the priority above all else.'

The result, he says, was no more winter stubble left on the fields for ground-nesting birds to lay their eggs in, while the development of new pesticides and herbicides, such as Glysophate, eradicated food sources. 'We had this magical substance which could kill all the weeds in the stubble, but which took away all the food for the wildlife. Obviously, nobody was thinking about it. It wasn't even a consideration.'

Shepherd, who is in his early 50s, is a keen conservationist and still practices a traditional mixed farming system, rotating around grass, on his land. He has helped establish a partnership with five other neighbouring landowners encompassing 7,000 acres, where they are monitoring the wildlife to try and protect some of the dwindling species. In his jacket pocket, he keeps a notebook to jot down the rare birds he sees while on his daily rounds: grey partridge, corn bunting and Montagu's harrier. A few years ago, a new species began to appear in his notes. Where so many other birds had been extirpated from the landscape through intensive agriculture, the raven has thrived with the constant supply of food all through the year, all over the Downs.

'We've always had hooded crows and rooks, but when the fields started turning black you knew the ravens had come. At first, I didn't mind having them and thought just another bird around the place. We were shooting the rooks – hundreds a year – and it made no difference whatsoever to the population. When the ravens took over, I thought this is bedlam.'

We drive down to the farm's southern fields where Shepherd points out to me the ravens' tactics. When the piglets are out of their pens, the raven on top of the shed drops down to block them from returning to safety, then another raven, waiting out of sight at the back of the shed, cuts around and they attack the piglet in tandem. 'They peck the back ends out of the piglets, and their eyes,' says Shepherd, who on bad weeks loses several piglets. 'They work in pairs. These animals are bloody clever.'

After watching the ravens lying in wait on his fields, we head for a thicket of Scots pine at the top of a hill known as Higher Boulsbury Wood, or unofficially by Shepherd as 'Raven HQ'. We hear the ravens before we see them, 'a primitive pterodactyl noise' as Shepherd puts it, before scrabbling up a steep slope to the top. We flush out a woodcock that hurtles away from us into some brambles and encounter several badger setts cratered into the side of the bank. Shepherd points out a centuries-old yew that he suspects may have been used for pagan worship, when from above us comes a similarly ancient sound.

This time the ravens don't fly off as we reach the summit, instead, six or seven of them stay on their perches watching us catch our breath as they murmur secret conversations. I spot a single massive nest at the top of one tree, but mainly this is a roost for juveniles to bide their time before making mad dashes back to the fields.

We explore the edges of the wood, ravens calling all the while above our heads, before Shepherd returns to his pigs. I stay up there for another hour or so, watching and listening to the birds that have evaded me for a week, immersed in their conversations and preoccupied by the catastrophic loss of wildlife Rob Shepherd has described; a result of the monocultures created by intensive farming, damage far worse than any one bird can wreak.

Perhaps, I wonder, the ravens' return represents a more visceral version of the destruction that we have wrought on the landscape but do not see for ourselves. I walk off down the hill with the ravens watching me go. Echoes of David Bowie are still in my ears as I dwell on thoughts of the fields turning black.

CHAPTER SIX

Bird of War

On 26 January 2016, a small procession of civic dignitaries weaved through the Caithness county town of Wick. A piper led the way, notes blasting off the rain-slicked pavement, marching the group towards the home of the Highland Council, Caithness House. Here, the Lord Lyon unveiled the new county flag to camera flashes from the local paparazzi and muted cheers from the assembled crowd, of the sort befitting a sodden Tuesday morning.

Caithness – a wind-blasted county on the northernmost point of Britain – is the first community on mainland Scotland to have gained its own flag. The winning design, which was decided upon following a public competition led by the *Caithness Courier* and *John O'Groats Journal*, features a Nordic Cross, in a nod to the Viking invaders that settled these distant shores. The flag is black, symbolizing the seams of locally quarried flagstone and the peat-rich bogs of the Flows, and embossed with gold and blue lines representing the jutting triangular coastline scored by the North Sea on one side, and the Atlantic on the other. In the first quarter sits a golden galley – the traditional emblem of Caithness – and on its sails is a raven.

These are harsh lands: prone to capricious squalls of foul weather, and ruled by birds of blood. Merlins and hen harriers cut low over the peat bogs in search of water voles and other prey, while in the summer they roam the river corridors flowing down to the north and east coast, whirling above the shimmering lochs and competing for dominance over the Flows. Even the bog plants here are in search of prey – carnivorous sundews trap midges and flies that stray too close to their sticky sap, ensnaring so many insects that individual plants can survive for up to 50 years on the nutrients.

Since forming at the end of the last Ice Age, the Flow Country – a rolling expanse of 400,000 hectares of peatland that stretches from Caithness to neighbouring Sutherland and includes the largest expanse of blanket bog in Europe – was left almost entirely unspoilt until 250 years ago when humans began to intervene.

The large-scale clearances of the eighteenth and nineteenth centuries led to the Highlands becoming one of the most sparsely populated parts of Europe. Tens of thousands of families emigrated or were expelled from the land and replaced, by and large, with sheep. The Highland Clearances were devised to fracture Gaelic culture and clan society and bring order and commerce to an area long deemed by the British establishment to be its most troublesome corner. Known as the 'white tide' (a reference to the incoming sheep) the land they left behind was radically altered.

Tenants were forcibly evicted and townships razed to be replaced by vast sheep farms. These were intended by the ruling landowners to feed and clothe the populations of Britain's burgeoning cities, and make them rich in the process. Before then, Highlanders grew small quantities of oats and *bere* (barley) in the glens, and in the summer, took their small herds of sheep, goats and cattle up to the *shieling*: the rough stone-built huts whose ruins still litter Scotland's mountain pastures like headstones. The clearances destroyed this way of life and led to commercial operations on a huge scale overseen by southern capitalist graziers.

Ravens, already resident over the Flows and ever the opportunist omnivores, thrived among the newly enlarged sheep farms and what became a steady supply of meat. While farmers could control a small herd of sheep kept close to their crofts, with livestock scattered all over the hills it was impossible to keep a close eye. Over the past century, plantations of conifers for commercial logging also began to spread across the Flows, providing the ideal roosting conditions for large mobs of juvenile ravens. In the wake of man's intervention upon the landscape the birds followed, and on to the very symbol of this county. I have come to Caithness to better understand the strained relationship between ravens and farmers. For here, savagery and persecution remain all too modern themes.

* * *

It is a foul, blizzard-swept morning during lambing season and Selena Swanson sits in the kitchen of her farmhouse. Her eyes are red raw through tiredness and something else again. She brews a cup of tea and slides a plate of biscuits towards me across the plastic-coated, red-and-white checked tablecloth. A washing machine clatters next to us as hail pips dash against the windows. The kitchen walls are adorned with medals of prize-winning animals among the Swanson herds of 110 cattle and 350 ewes (pronounced *yaws* in the local dialect) alongside pictures of Selena and her husband John's two children, Bethany and James. Her son, in particular, Selena explains, is 'sheep daft' and already at the age of eight tends a few of his own. Her husband, who until that moment has been sitting silently at the table in a woolly hat and fleece sipping his tea, suddenly interjects: 'With them being so young I think they assume going out and picking up dead lambs every day is normal,' he says.

Framside Farm, near the small Caithness town of Calder, has been in the family since 1986. John and Selena both trace ancestors going back several generations in the area. A decade ago, they were lambing as sheep farmers here always have, keeping their herd out in the open throughout the season. This allows the ewes to give birth naturally in the fields and hardens the lambs from a younger age, thus ensuring against the spread of infectious diseases such as navel ill – a form of bacterial arthritis that spreads easily across the straw mattresses of the indoor sheds.

Selena says there had been problems with the ravens before, but they managed all right. Raven numbers in Scotland have rocketed to around 6,000 breeding pairs, and juvenile flocks, several dozen strong, now occupy single fields in Caithness wreaking havoc among the sheep. The size and ferocity of the raven mobs have forced the Swansons to start lambing in sheds instead. Nowadays the newborns are kept inside for at least three days before they are released on to the fields; even still, the birds are waiting for them. 'You don't count how many you lose,' Selena says, softly, tears welling up in her eyes again. 'You would be suicidal if you did. When the ravens see you putting the lambs out they gang up. When the lamb is susceptible, and they have no need to go in with more than one animal, then they see the opportunity just lying in front of them, and go in one bird at a time. But when animals are up on their feet and running around a field, that is when you notice them gathering in twos and threes and trying to split the lamb from its mother or even attack the lamb. I've seen one raven on a lamb's back and the other holding on to its tail. And that's a perfectly fit lamb running around a field, but it has no hope. They're preying on the lambs.'

Selena keeps North Country Cheviots, one of the new breeds that appeared during the aristocratic experiments of the Highland Clearances. They were introduced to northern Scotland by Sir John Sinclair, a member of the Earls of Caithness dynasty. His fifteenth-century ancestors turned the now ruined Ravenscraig Castle on the Firth of Forth into what was at the time one of the most well defended forts in Scotland. It was also the castle upon which Sir Walter Scott's ballad *Rosabelle* was set, in which a young girl makes the fateful decision to attempt to cross the water on a stormy day and drowns. There is a line in that poem which reminds me of the ravens in the lambing fields: 'The blackening wave is edged with white … Whose screams forebode that wreck is nigh.'

Sir John Sinclair of Ulbster was a politician and agricultural pioneer who founded the Society for the Improvement of British Wool. In his Caithness constituency, he had started attempting to mix new breeds of sheep from around 1785, and in 1788 bought the Langwell estate, where he introduced a herd of 500 Cheviot ewes from the Scottish Borders. According to the academic and expert in agricultural history, Dr M. L. Ryder, in *Sheep and the Clearances in the Scottish Highlands*, the Cheviot – known locally as 'long sheep' – evolved almost entirely from the native Dunface or Old Scottish Shortwool. Following Sinclair's introduction of the sheep into Caithness and Sutherland, the breed evolved into the North Country Cheviot. The animal, so Ryder says, was regarded with particular bitterness and hated by Highlanders, who had been forced from their glens to accommodate its prolific winter grazing. As the centuries have slipped past, that hatred is being transferred to the ravens, a new interloper upon these lands.

At the Swanson farm, the North Country Cheviots always lamb first in the season, followed by the less pedigree Texel cross. The raven flocks begin to build earlier in the year and by spring they easily number 20 or so on each field. 'I've lost two to three lambs a day this year,' Selena says. 'Yesterday I lost five. I nearly caught one of the ravens a few days ago. I had a mother with three lambs out in the field and it was eating the dead lamb right next to her. You don't want to get out of your bed any more because you know your lambs have been cowering in some corner. We've lost thousands of pounds, and it's a threat to our livelihood in a sense, but it's more than that. We've grown up with farming. It's in our blood.'

Through the course of my research I have encountered a curious disconnect when it comes to raven attacks on livestock. Many lovers of the bird have outright refused to accept that ravens ever predate healthy farm animals. Their argument is that the animals are already dead or dying – brought down by a fox, lameness or simply the cold of an unseasonably spring day, like the lamb I met at the Swansons – and the ravens simply set upon the corpse. The farmer arrives and finds the birds pecking out the eyes and haunches and pins the blame on the raven. But this contradicts many of the historic accounts I have read and what I have seen with my own eyes. As opportunists clever enough to plan and carry out attacks in partnership – as I had already witnessed on the pig farms of the New Forest – why would the ravens ignore the easy meat of a newborn lamb?

When we finish our tea, Selena leads me to the lambing shed opposite the house. Outside, arranged in a pathetic huddle, are the three corpses of lambs she had discovered the previous day. Each tiny lamb has had its eyes and tongue pecked out, leaving livid slits in the matted fur, and has been partially disembowelled. The ravens go for the eyes and back ends because that is where the meat is softest and because they are the most difficult places to defend. When the intestine is exposed, the birds then draw it out. I instinctively recoil, but

Selena trudges wearily past. Inside the shed another lamb, only a few days old, shivers close against its mother – often, the crofters say, a surviving lamb that has been mauled by ravens is rejected by its parent because it carries the signs and smell of the bird. The lamb has blue paint daubed on its side indicating when it was born, and when it does move it is with an unsteady limp. Selena scoops it up bleating and shows me the stab wounds in its fur and the bloodied stump where its tail should be, covered with a yellow stain of iodine. The tough shepherds of Caithness do not often name the individual animals in their herds. But Selena has taken to calling this one Stumpy – notable for being one of the few to have survived a raven attack.

* * *

We climb into her 4x4 and take a drive over the farmland to assess that day's casualties. There are four main fields surrounding the Swanson farm and on a fine morning sunlight picks out endless colours in the landscape. The pastures can appear so green and the sea so blue, it's as though they've been drawn in child's crayon, while the grey flagstones that mark the field boundaries are interspersed with blooms of deep red and purple, like clouds curdling across a summer's full moon. Not today, though. Today, the thunder clouds blowing in have muddied the palette and jumbled sea, sky and land. The sheep stand out yellow against the dusting of snow pips swept about by the squalls. The ravens are Bible black.

The fierce winds that batter this part of Scotland mean that instead of opting for hedgerows or drystone walls to border fields, farmers use slabs of Caithness flagstone, formed in the Devonian period some 400 million years ago. The stones underline the ground in seams up to 4km (2.5 miles) thick and their layered structure allows them to be easily split.

Neolithic stonemasons used Caithness flagstone to build the brochs and burial mounds that still dot the countryside. These include the chambered tombs of the Grey Cairns of Camster, built 5,500 years ago on a windswept moor not far from the Swanson farm in the heart of the Flow Country. By the nineteenth century, the flagstones were being shipped all over the world, paving booming capitals such as Sydney in Australia and Montevideo in Uruguay. The stone remains sought after today, and it is Caithness flag upon which politicians' footprints ring out in the new Scottish Parliament building.

A quarry still quietly churns next door to the Swanson farm, its blackened heaps of rocks looming like slag heaps over the fields. It is mid morning and the ravens have already assumed their positions perched on the flagstone fence boundaries overlooking the lambs. Columns of hooded crows march among their larger cousins about the pasture, hooking out earthworms that rise too close to the damp grass. The ravens appear to be juveniles – at a guess two to three years old – happy to hunt in flocks if necessary and roost together at night among the conifers on the brow of the hill above us. On my drive to the farm a few hours earlier, I had seen one raven lurch off the flagstone and down towards a pair of lambs. The bird missed its strike and the lamb it was aiming for bolted behind its mother just in time, but in that brief flurry of attack the raven assumed a nightmarish shape: wings spread and arched, beak partly opened and stabbing down. Black and white jarring on the Caithness Flows.

The windscreen wiper on the 4x4 works furiously as we bump across the ground, steamed-up windows further obscuring the view. We spot ravens in distant fields, but each time we enter through a gate the birds take off and disperse until we have left. Similarly, each time we wind down a window to clear the steam, any raven within a few hundred metres rises and disappears into the gloom. John, who is driving, says the birds have learnt their daily movements about the farm and take flight at the hint of anything out of the ordinary. 'The boldness and intelligence of them is hard to watch,' he says. 'The kids have made scarecrows but it had no impact whatsoever. For a few days, we set off bird-scaring bangers, which drove them to the opposite end of the fields but as soon as we stopped they were back. I found one lamb a few days ago with a raven on it, and rescued it and took it back to the house to warm up. I got a bottle of milk for it and tried to get it to feed, but realised it didn't have a tongue any more.'

As we sit, wipers whirring, Selena points to a corner beneath the flagstone fence where the previous year she found 10 lambs killed in one single morning by what she says were ravens. 'Anybody that wants to, should come out here and see what's happening and tell me that's not cruel,' she says.

* * *

I have come to Caithness with a man called Ian Rutherford, a photographer and bird enthusiast who grew up on a Scottish croft. In spite of his agricultural heritage, farmers in this close-knit community are normally reluctant to talk to people like us turning up with cameras and notebooks and pens. Most are still smarting from a *News of the World* journalist who came up here a few years ago on the trail of a big cat on the loose. The story was silly season stuff; the product of fevered Fleet Street news desks' imaginations, and the reporter in question supposedly misquoted all whom he interviewed. That legacy of mistrust lives on.

We, however, have a way in. A few weeks before our visit, a petition was started calling on ravens to be added to the general licence. It was arranged by Danny Bisset, an oil rig worker, shooting enthusiast and amateur pest controller, who originally hails from the Black Isle north of Inverness, but married into these lands and lives with his family in the village of Reay. A short, portly man in his 30s, Danny has a huntsman's view of the countryside; he speaks of many animals as if he were viewing them down the sight of a rifle and in defence of his activities is a member of a local shooting syndicate. He launched his petition to be allowed to do to ravens what he already does to crows on farms: trap and shoot them en masse, and has urged me to come and see for myself the damage they do.

Ravens are currently protected under the Wildlife and Countryside Act of 1981, which prevents their killing unless a special licence has been granted. Already in 2009, the numbers permitted to be shot under licence in Scotland was increased by Scottish Natural Heritage, but Bisset's petition is calling for another step altogether: to add the raven to the general licence, enabling farmers to kill any raven worrying their livestock or indeed any of the birds that takes their fancy. The present system that typically only allows licence holders to kill a pair of ravens at any one time, makes no difference whatsoever according to Bisset and his supporters. By the time we meet, his petition has received more than 2,300 signatories. However, a counter-petition launched in response, has so far been signed more than 27,000 times, warning that unregulated killing could again lead to the raven becoming decimated. At this discrepancy, Bisset shrugs, and points to the global addresses of those who have signed the counter-petition online. This is a local problem, he says, one outsiders cannot comprehend.

Strangely enough, for a man campaigning to kill ravens at will, he is coy about the exact methods of dispatching the birds. Only after several attempts at questioning does he relent, and shows me a crow trap on one local farm; a small wire aviary where the birds are lured in and then blasted at close range with a shotgun. Bisset has worked for several weeks to persuade farmers beleaguered by ravens to talk to me and show me the damage caused by the birds at first hand. As we tour the various areas that are being worst hit, it is fascinating to hear the language invoked by the crofters and the similarities it bears to the century-old myths about the birds. Ravens are not discussed as simple predators, but in deeper, more moralistic terms.

At Achaquil Farm in Scotscalder, overlooking a loch and the mountains that tower over the Flows, Hamish Ritchie, a 65-year-old sheep farmer who has been working on the land his entire life, describes to me the sight of three ravens with their wings spread out, slowly herding one of his lambs into the corner of a field. He was watching from a distance, too far off to scare the birds away, and says the job was done so expertly that at first he thought it was a wayward working dog. 'I watched it for as long as I could until the lamb went down,' he says. 'It is hellish to see. You can't turn your back for a second.'

When we meet, Hamish is in the process of applying for a licence to shoot ravens. When I ask whether in such an isolated area the farmers just shoot the birds regardless of officialdom, he tells me an apocryphal story of a fellow crofter who lived on the Isle of Mull. The man there had grown weary of ravens setting upon his flock so took up his shotgun and blasted one out of the sky. The shooting was reported straight away and the farmer learned that two policemen were coming to investigate. When they arrived, there was no sign of the bird or weapon, only a large cooking pot simmering on the stove. 'How did it taste?' 'How do you think,' laughs Hamish. 'But he got away with it.'

Hamish's wife, Les, says she has even seen ravens circling over their two sheepdogs while out walking on the moors.

'Ravens are really different to other birds. I'm too much of a lady to say what I actually think of them, but I would say they are evil. They're worse than vultures for the simple reason that a vulture will wait for its meal to pop its clogs, but a raven won't. To see a lamb or ewe being attacked alive is horrendous. There is nothing you can do.'

That ferocious intelligence, the sense that the birds are watching the farmer's every move, comes out in conversation again and again. Each farmer I speak with begins by making the economic argument against ravens killing their lambs, but soon this descends into a deeper loathing: a claim to be once more under siege on their own lands. Nobody, though, speaks of the possibility of diverting the birds through other means. Only of meeting savagery with savagery.

* * *

Where the country falls into the sea, the cliffs turn to rust.
The Old Red Sandstone face of Dunnet Head represents the
most northerly point on the British mainland. Battered into
ever sharpening right angles by the waters of the Pentland
Firth, as the cliffs recede the foam flecks tease out new
cracks and fissures to be dissolved by the waves. Dunnet
Head is the sort of outpost that inevitably becomes the site of
last defence.

In 1940 Vice Admiral Sir James Somerville chose this
site as the last link in a radar network known as the
Admiralty Experimental Stations. Dunnet Head was
renamed in official naval parlance as A.E.S. No. 6, following
those already built at Sumburgh, Saxavord, South Ronaldsay
and two on Fair Isle. It became operational in December
1940 and remained in use throughout the war, tracking
aircraft up to 160km (99 miles) away and U-boats breaching
the surface far closer to the mainland. At first the station
consisted of two utilitarian brick huts, the receiver in one
and transmitter in the other. According to local historians
the aerials were turned by hand using the cranks of upturned
bicycle frames.

The remnants of military occupation remain among the
tufting moorland clifftops. In spring when I visit, banks of
purple Scottish primrose that grows only in a few select
places in Scotland, have bloomed into bursts of star-shaped
purple all over the precipitous heath. The Mackay family
farm sits a mile or so away inland. Looking out towards
Dunnet Head, Andrew Mackay, who runs the business in a
three-man operation with his older brother Joe and father,
Andrew Senior, can watch the ravens heading back towards
their roost on the cliff edge at night. Mackay was the first
man in the area to be granted a licence to kill ravens and
reckons he has probably shot more than anybody else in
Caithness. Taking off his yellow baseball cap and scratching

his sun-weathered head, he estimates he has killed 30–40 ravens in the past six years or so. A .243-calibre rifle is normally his weapon of choice.

'It's pure luck,' he says. 'You've got to take them at a distance because you can't get near enough one normally. If the opportunity comes up you can get them with a shotgun. I was shooting ducks one night and one flew over. But that's luck. If you want to pick them in a field you need a big, high-calibre rifle, because it's 300 yards distance needed between you and the bird.' A few years ago, Mackay noticed the ravens were starting to gather on an old truck in the middle of one of his fields so, early one morning, he headed out there and sat inside with his shotgun primed. 'I lay on the back of the truck and they wouldn't come within 100 yards of me,' he says. 'The next day though they were back on the truck. Clever bloody things.'

Previously, he says, he would get a licence to shoot two birds and then immediately apply for another one at the Scottish Rural Affairs (later Scottish Natural Heritage) office in Thurso. Each licence took from a couple of days to a week to come through. Two years ago though, he was refused another licence. 'They told me there was no need for a licence, they were now under control in my area and that was it. They said you've shot so many there are no more.'

This season, Mackay estimates, the family has lost more than 60 out of 900–1,000 lambs and 10–12 ewes to ravens. 'When I was a kid we didn't even know what a raven was – just something in the Tower of London – and even 10 years ago they weren't here. We don't want them wiped out. Just brought under control. There's no coming back from these raven attacks. Unless you get there in time the lamb they've set upon is just destroyed. When you shoot one it scares them for a wee while, but they just come back. They see it as easy food.'

Aside from the attacks upon his flock, Mackay raises another issue with the growing number of ravens in the area: the impact upon ground-nesting birds. He estimates they used to have about 30–40 pairs of lapwing on their fields, but nowadays this has reduced to five. Earlier that morning at the Swanson farm, when I was crouched by one of the flagstones watching the ravens swirl about the field, I noticed one swoop down upon a lapwing nest. The raven made four or five attempts to snatch a chick, and each time, the far smaller lapwing burst up to meet the aggressor, shrieking its warning cry like a barrage of electrical bleeps. From where I was watching, the raven's attack appeared to be unsuccessful. I breathed a sigh of relief for the valiant peewit.

Motoring in Bisset's 4x4 a few miles away from the Mackay farm we see a curlew shoot up into the sky locked into a duel with a raven. The wader swishes about its curved beak like a scimitar to deter the raven, which is aiming towards its nest – just out of sight behind the road verge. As ever, when being harassed by another bird, the raven simply coasts alongside its rival tipping its wings to shirk from the blows with consummate ease. We leave before seeing whether the curlew succeeded in keeping the raven at bay.

The rise of the raven and decline of waders like the lapwing, curlew, golden plover, snipe and dunlin in upland areas, whose numbers have more than halved since 1985, seems to be an obvious equation. But in 2010 a joint study between the RSPB and the University of Aberdeen analysed data from $1,700km^2$ (656 square miles) of Britain's uplands, including Exmoor, the Lake District and the Scottish Highlands, to explore the patterns of change in ravens and wader abundance over the previous 20 years. The study found only 'weak associations' between the increase in ravens and the decline of lapwing and curlew, and concluded that there was little evidence to show waders had decreased because of the growing number of ravens. Instead, it implied that other factors, including changes to habitat and vegetation cover, as well as a general increase in other predators like foxes could be responsible for these large-scale changes. Another predation study published in 2017, after being commissioned by the Scottish Government, warned more data was required to accurately assess the true impact of ravens on ground-nesting birds.

We drive a few miles away, with the mountains of Caithness looming in the distance, towards the small town of Thurso and stop by a row of old cottages just off the main road. Inside, I find myself sitting at another farmhouse kitchen table opposite two brothers, in a house attached to a hill farm they keep with an uncle a few miles away. I am leaving their names out because of a particular story one of them let slip as we sip our tea. The family, he tells me in gruff, staccato Scots, has been sheep farming since 1966 and their herd has grown to several hundred breeding ewes. Lambing always took place out in the open on the fields, with no problems with ravens whatsoever, then about 10 years ago the birds started coming. The worst year was in 2010 when the family lost about 100 lambs. The attacks came from a new commercial conifer plantation that had been established above the farm and had since turned into a large raven roost. 'We took someone up to have a look and he said it was a young male colony,' the farmer says. 'He said, "you've got a lot of testosterone up here".'

After failing to persuade the RSPB that they had a problem, the farmer says he was approved for a licence and one evening went up to the roost with a shotgun and took aim at the nearest bird he could find. The body of one raven fell to the forest floor. Around its foot was a metal tag that read: 'Tower of London, property of Her Majesty the Queen.' He has barely finished the tale before admitting: 'I probably shouldn't have told you that bit.'

In an era before environmental protection, the farmers in Caithness used to relish the annual 'corbie shoots'. Families would gather at a particular farm, load up their shotguns with fresh cartridges and head to a nearby roost where they would shoot indiscriminately into the trees. Crows, rooks, ravens and hooded crows and their young would fall to the floor in their hundreds; another generation wiped out for the year. The older farmers here speak of these shoots with a wistful nostalgia. It is no surprise that most of the younger Caithness crofters grew up never setting eyes on a raven; as they had been blasted to the county's furthest reaches, clinging on to the blood-red sea cliffs.

'Obviously there are people out there talking about total annihilation,' the farmer says. 'We don't want them to be made extinct, just controlled. Their intelligence is impressive, no doubt, but from a farmer's point of view, there is nothing that would make me want to have them hanging about.'

* * *

In the 1940s, the Swiss zoologist Adolf Portmann who had developed his pioneering work at the University of Basel, undertook a study to measure the brains of hundreds of species of bird. He concluded that corvids, in general, have larger brains relative to their body size than all other birds, with ravens top of his list. The largest brain size found in a common raven is 15.4g (.5oz) and the size of a large Brazil nut, compared to an average body weight of about 1kg (2lb). For much of the twentieth century, the size of the brain was regarded by scientists as the factor of primary importance in determining an animal's intelligence, but new studies have challenged this. In June 2016 a team of international researchers from Vienna, Prague and Brazil published research demonstrating that birds have a vast amount of neurons densely packed into their brains, enabling some species with an intelligence comparable to, or even exceeding, primates.

Ravens, along with parrot species such as the blue-and-yellow macaw and sulphur-crested cockatoo, came out on top. The raven used in the study had a brain mass of 14.4g within which was contained 2.171 billion neurons overall. Of these, 1.204 billion neurons were found in the forebrain, a much higher concentration compared with primates, or other mammals and birds. Despite lacking the cerebral cortex upon which humans rely as the nucleus of our intelligence, in the 1960s neurologist Stanley Cobb found that the forebrain in birds gives them the capacity to make complex decisions. For ravens – and other corvids – this means they can manufacture and use tools, solve problems, develop complex social networks, learn to mimic voices, recognise themselves in a mirror, and also recognise human faces.

In 2008, the US wildlife biologist, and authority on crows and ravens, John Marzluff released the first formal study of facial recognition in wild birds. It proved that corvids could identify individual humans – something the farmers of Caithness have long claimed. Dr Marzluff and two of his students trapped seven crows on a university campus in Seattle, while wearing rubber caveman masks. After the birds were released they walked about the campus wearing the caveman mask and a 'neutral' visage of the former US Vice President Dick Cheney. When 'Cheney' strolled past, the crows remained calm, but whenever the caveman appeared the crows would scold whoever was wearing it. As the years progressed the reaction to the mask became increasingly severe. When the study ended after two years, Dr Marzluff had been scolded by 47 out of the 53 crows he'd encountered while wearing the caveman mask. They concluded that the birds learned to identify threatening humans from parents and others in the flock.

Another wildlife researcher, Stacia Backensto, has been developing this work. Backensto has been studying the raven population in the oil fields on Alaska's North Slope and the impact of human intervention on birds in wild landscapes. The ravens first began to appear around the oilfield camps in winter a few years ago and their population has increased with the development of heavy industry on the North Slope. Right now, more than 100 ravens arrive in the dead of winter when their summer diet of lemming, shrew, and eider goose and duck eggs dries up. They build their nests on the drilling rigs and communication towers out of wire and plastic cable ties. Backensto was so convinced that the ravens she had previously tagged remembered her that she donned Carhartt overalls, a black cap and a fake moustache to resemble an oil worker and fool the birds into allowing her approach. Her disguise worked.

In Caithness, Ian and I are warned straight away by the farmers that as soon as we appear on the fields the ravens will take flight at the sight of two new figures, and so it proves. The ever-present threat of being shot means the ravens are hyper vigilant. While they tolerate me watching them through my binoculars, the moment Ian produces a long lens camera the birds scatter from the sheep pastures. That said, the ravens never flew with any sense of urgency, as if establishing dominance over us interlopers even when taking flight.

After a failed day of trying to get close enough to ravens to photograph them alongside lambs, Ian and I meet another farmer and hatch a plan. Helen Forbes runs her own 190-acre farm single-handedly each week while her husband works for an oil company in Aberdeen. This year she has lambed 180 ewes and wherever she goes she is tailed by her sheepdog Bob, who growls low to the ground at the presence of another male on the farm, situated on land which lies close to a large raven roost. Whenever a lamb has been brought down or she scatters food pellets for her flock, she hears a loud cry started up by the ravens and more and more birds mob over. 'When you hear the ravens' feeding call it just sounds like they are laughing at you,' she says.

John Marzluff and Bernd Heinrich, another US professor and raven expert who has written numerous books on the bird, employ a different phrase for it. Having analysed raven behaviour in roosts in the great forests of New England as well as the Welsh countryside, they call roosts 'mobile information centres', where knowledge of rich food patches is quickly passed through social groups. After locating a food source, the birds call or soar above it to attract others and achieve dominance over the feed. At dawn, the ravens leave their roosts to hunt and forage in 'highly synchronised groups'.

Helen Forbes normally anticipates the arrival of the birds on her farm at around 5.30 am when she goes out to feed her sheep. Each morning she and Bob go out on the quad bike to scatter pellets of grain among the sheep in the fields. When the ravens see the food being dropped, one will cry out and within minutes a flock will descend to eat among the sheep. Normally, they do not bother to attack the livestock when there are pellets to be scooped up instead.

We know from our experience visiting the other farms, that if we try and watch the morning feed the ravens will recognise our strange forms in the field and stay away. Instead, we construct a bird hide in the corner of the field overnight – when the ravens are back in their roost and cannot see what we are doing. The next morning, when Helen heads out on her quad bike, we sit on the back, jump off when it passes the hide and scramble in as quickly as possible. It is a plan executed with military efficiency. As the quad bike putters off, we are left peeking out of spy holes only 6 or so metres from where the ewes and their lambs are feeding. Just as Helen describes, the ravens' feeding cry goes up – a distinct and higher timbre than its more typical croak – and the birds begin to descend. A flock of a dozen or so line up on fence posts where they can pounce upon the food, rasping out their raven refrain through their beards. Others wobble across the grass from claw to claw, moving like prisoners in chains.

Being this close to the birds the size is what strikes you straight away: some of the larger ravens are of equivalent size to the lambs. It is a beautiful clear morning and as they arrive en masse it is possible to see feathers missing in their taut wings and fanned tails, which look like a roughly drawn ace of spades. The ravens' feet are ringed with silver hoops, and curved claws, a good few centimetres long, drape over the fence posts upon which they perch. While none of them makes a lunge for the lambs, preferring the easier meal of pellets dotted about them, an air of menace pervades the scene. The sheep occasionally bleat but the ravens, once they have landed, are by and large silent. 'Click', goes Ian's camera, as the birds blink their eyelids closed like steel shutters. Occasionally they glance towards our hide and keep a distance, but we do not give our position away. The feed continues for an hour or so, before the ravens take flight towards a perch on the distant fields overlooking the rest of Helen's herd. With breakfast over, the ravens turn their minds to the other meals of the day – and Helen dreads what they will choose.

My final night in Caithness, I plan with Danny Bisset to try and visit one of the raven roosts in another newly planted commercial forest on what was previously the open land of the Flows. We meet long after dark, with the lamps on the top of his 4x4 fizzing in the misty rain. We drive for a while then pull into a lay-by. Danny points to the outline of a pine forest in the distance where the ravens sleep.

We begin walking towards the roost but realise we are heading straight through a peat bog. With each step, our boots sink down to the calf, or deeper. We had set off at a quick pace, with Danny telling me some of his shooting stories, but our laboured efforts soon sap our energy and we fall silent, apart from our rasping breath. The ground sucks and pulls us down, as if resisting any more human footprints upon it. Eventually, with sweat pouring down our faces and about a quarter of a mile still to go to reach the forest, we give up and turn back. No doubt the ravens see us approach and then retreat. Two strange figures stumbling and cursing in the bog.

Long after I have left Caithness, this battle between ravens and farmers stays in my thoughts. It is clear that human changes to the landscape have created the ideal conditions for juvenile flocks to gather. The commercial forests create the perfect places for them to roost, while the expansion of pastureland and management of the countryside, which has led to the decline in other birds like waders, has reduced their available food, focusing their attention on the lambs and bringing them into closer conflict with humans. That explanation is no solace to the farmers of Caithness to whom anybody would have the utmost sympathy for picking lamb cadavers off their fields each day. Perhaps providing financial compensation – as Scottish Natural Heritage have offered to farmers on the Isle of Mull whose lambs are attacked by the burgeoning white-tailed eagle population – might work, or perhaps what is needed is a whole new emphasis on the management of the land? Surely a bird as intelligent as the raven can be coaxed into a more peaceful coexistence.

A year after my visit, Scottish National Heritage began a new trial to address the problems between ravens and farmers in Caithness, extending the licence to permit killing the birds collectively across a number of farms, rather than just as individuals. Their aim is to appease the suffering farmers and build up a far clearer idea of where the attacks are taking place.

While addressing the mistrust between conservationists and farmers by amassing exact evidence seems welcome and long overdue, I still feel deflated by this new arrangement. I have no particular issue with culls if a dominant species is threatening biodiversity, but this is not the case in Caithness – or at least has not yet been proven with regard to the impact of ravens on waders. Rather this is all about our own reactions to ravens. Conflict, like a landscape, is something that we create. We decide which animals are allowed to kill in peace and which are not.

I arrange to speak on the phone with a man called Davy McCracken, a professor of agricultural ecology, and expert in managing conflict in the Highlands. I put to him some possible solutions that do not involve a rifle, and he gently picks apart each one in turn. Compensation? According to Davey, the sea eagle scheme is fraught with the difficulty of actually proving what killed an individual animal, and would be unfeasible with the scale of the raven attacks. Diversionary feeding? He thinks this may actually serve to maintain, or even increase, juvenile populations of ravens and before doing so we need far better research into the formation and motivation of the young birds. Restoring the landscape to how it was before being adapted for forestry and large-scale sheep farming could be a solution, he admits, but one that would take decades to come to fruition. And then, he asks, what exactly would it be restored to?

'We don't have any wilderness areas left in Scotland,' he says. 'All of this land has been managed by man in some way or another for thousands of years. Who really knows what "nature" is? These conflicts remain exactly that because there are no easy answers.'

When I embarked on this book I expected to encounter polarising views of ravens in my travels around the country, but I was still taken aback by the strength of feeling in Caithness. For ravens to feature on the county flag while at the same time its farmers campaign to shoot them at will, is a strange juxtaposition. I find it impossible not to feel conflicted about the return of the raven here. And yet there are other parts of the country, even close by, where the birds are revered. Before my visit to Caithness, I had already hoped to cross the Pentland Firth to the Orkney Islands and find out more about the relationship between people and ravens in a place where Norse culture remains so entrenched and celebrated. My time with the men who shoot ravens has made up my mind.

CHAPTER SEVEN
The Viking Survivors

'What a howling I've caused: the corbie croaks over
carrion in Orkney.' Earl Einar after carving the blood eagle
into the back of Halfdan Longleg as a sacrifice to Odin.

Orkneyinga Saga

The sun is shining through the narrow, centuries-old, stone-framed windows of the farmhouse where Fran Flett Hollinrake sits, casting a half-shadow on her face in the manner of Rembrandt. It is high summer but a windy day, as they almost always are in the hills of the West Mainland, Orkney. The wood-burning stove, which has not yet been lit that morning, gives off a metallic hum with each gust that is sucked down the chimney.

Reclining on her sofa, she cradles a cup of tea and a plate bearing a scone buttered with home-made jam, made from the stalks of wild rhubarb whirling about like sea anemones in the garden outside. Two of her three 'semi-feral' cats – one possessing only a single good eye, and the other convulsing in perpetual fits of sneezing – prowl the floorboards hoping for a crumb. Despite Fran enthusiastically waving her arms and shaking her curly hair, as she conjures the stories of the old Viking earls of Orkney, the cats wait in vain.

She is telling me about Sigurd the Stout, one of the earls who ruled over Orkney from the ninth to the thirteenth centuries following the conquests of the islands by the great kings of Norway, and the story of his famous raven banner. According to the sagas, Sigurd was a feared chieftain and powerful ruler, able to defend his dominions on the Scottish mainland, as well as launching plundering raids each summer across the Hebrides, Scotland and Ireland. His empire slowly expanded as he established an iron grip over northernmost Britain, until one year, Sigurd was challenged by a rival Scottish earl to meet his army in battle at Skitten in Caithness.

Fearing defeat and reckoning the odds against him were at least seven to one, Sigurd went to his mother, a sorceress, and asked her to conjure a spell to ensure victory. Her response was typical of a blood-soaked age: 'Had I thought you might live for ever,' Fran recounts to me, leaning forward and fixing my eye, 'I'd have reared you in my wool basket. But lifetimes are shaped by what will be, not by where you are.' Instead, she wove him a banner decorated with an image of a raven, embroidered so skilfully that when it fluttered in the wind, the bird appeared to be flying ahead of whoever held it aloft. 'It will bring victory to the man it's carried before,' Sigurd's mother told him, 'but death to the one who carries it.'

The moment the two armies clashed at Skitten, Sigurd's standard-bearer was instantly struck dead. Sigurd ordered another of his warriors to pick the banner up in his place, and he too was killed. By the time the battle was over, the corpses of three standard-bearers lay among the dead, but Sigurd was victorious. A few years later on Good Friday 1014, Earl Sigurd was required to raise his raven banner once more in the Battle of Clontarf against King Brian Boru of Ireland. Knowing this time what would certainly befall them, each warrior among Sigurd's ranks refused to carry the banner, so instead, he was forced to wield it himself. He carried the raven with a roar into battle but did not last long before he, too, met his inevitable fate.

First written in Old Norse in the *Orkneyinga Sagas* of around 1200 by an unknown author, Sigurd's story has seeped into Orcadian folklore. Until the eighteenth century, these fireside stories were regularly told in homes all over the islands, before the church began to outlaw such tales of magic and barbarism. The stories in the sagas are savage beyond modern comprehension. In the tale of the *Blood Eagle*, a victim's back is split open with a sword, his ribs pulled out, and lungs draped over his shoulders – a means of dispatch much celebrated by the old chroniclers. So too, the act of locking your rival and his men in his drinking hall and burning them alive.

Another Orcadian tale from *Njál's Saga* (a 13th Icelandic Saga composed around the same time as the *Orkneyinga Sagas*), goes that when a great Irish battle was raging, a man stumbled across a group of strange women working by night in a weaving shed. When he looked through a slit in the window, he found a loom set up with spears. Rather than wool, they weaved human entrails weighed down with human heads and used a sword as the shuttle, and arrows for reels. These 'weird sisters' as the saga calls them, were the Valkyries or 'the choosers of the slain', deciding who would die in battle. 'They are stories of absolute horror, but culturally people here are very proud of their Viking identity,' Fran tells me. 'Viking culture was so strong it obliterated what went before. It is very distinct.'

The Flett in her surname is a sure sign of her Viking lineage. It is a common enough name in Orkney, often painted above the shop windows in the two main towns; Kirkwall and Stromness, as well as bulking out pages of the phone book. Fran was born in Elgin and only traced her roots back to Orkney later in life. Before moving back to live on the island some years ago, she visited Orkney with her husband, Andrew, on their honeymoon. A few days into their trip, he pointed out a woman in a chip shop who closely resembled her. Even now she comes across her doppelgangers modelling Orcadian jewellery and Northern Isles tweed in the local magazines. 'There are quite a few people here who do look like me,' she admits. 'One thing the Fletts have are these big dark eyebrows.'

Fran, who in 1991 became one of the first four people in the world to graduate from university with a degree in Scottish history, traces the name back to one man, Thorkel Flettir. He was a Viking who lived on the island of Westray, a rugged northern outcrop of the Orcadian archipelago, whose few hundred inhabitants are known for their ancient DNA that predates even the Norsemen. 'Somewhere along the line Flettir got shortened, but from him are said to be descended all the Fletts in the world,' she says. 'He was a very violent man. There is a bit of debate about what Flettir means, but the thing most people have gone for is flayer.'

* * *

A few years ago, genealogical researchers compiled 5,000 samples from men across Britain, tracing their past through the Y-chromosome DNA that is passed from father to son. Men were used in the project, as historically women moved around more as marriage often required them to leave their family homes. According to the research, 29.2 per cent of descendants in Shetland were found to possess Viking DNA, 25.2 per cent in Orkney and 17.5 per cent in Caithness. By contrast, just 5.6 per cent of men tested in Yorkshire, where the Viking kings also ruled, were found to possess the same DNA. It is known in genetics as the M17 marker, the true signature of Viking blood. When researchers narrowed their findings to men with ancient Orcadian surnames like Flett, Linklater, Foubister, Sinclair, Clouston, or Rendall, the percentage of M17 rocketed to 75 per cent. Even the rodents in Orkney can lay claim to such rare blood. Researchers conducted a series of tests on house mice in Orkney, and discovered that their genetic make-up was utterly distinct from those on the Scottish mainland. When they compared the Orcadian samples with those taken from mice in Norway, they discovered an exact match.

Of mice, of men, of buildings, of blood: the old Norse ways still define these islands. Its capital, Kirkwall, remains dominated by the vast St Magnus Cathedral founded in 1137 by the Viking Earl Rognvald in honour of his uncle. Fran Flett Hollinrake, who is custodian of the cathedral, says the Flett family crest is embossed on one of its stained-glass windows, and while at work she regularly meets descendants from across the world who have come to pay homage to their Viking ancestry. Streets and villages still have Old Norse names, and the raven, the very emblem of the Norsemen, still dominates its towering clifftops, culture and folklore. In recent years, they have been spotted circling about the cathedral spires once more, looking for a place to build a nest.

Along with the wolf and the eagle, the raven was revered as a sacred animal by the Vikings and remains an ever-present symbol of Orkney today where bottles of Raven Ale are for sale on the Kirkwall High Street. The great Norse god Odin carried two ravens with him: Huginn (thought) and Muninn (mind) who would fly off to distant realms and return whispering eternal knowledge and all-seeing secrets into his ears. In the sagas and skaldic verse (the praise poetry commissioned by Viking rulers to tell of their own great deeds) the birds are almost a constant presence. In Old Norse, ravens are called 'corbies', 'the taster of the corpse sea' and warriors known as 'foot-reddeners' in anticipation of the birds clambering across their fallen bodies.

As in Old English poetry, the Norse sagas used ravens as a precursor to death; in battle the birds supposedly stalked those who were going to fall. Another Orkney earl, Thorfinn the Mighty, one of the most powerful rulers of the islands, was described by his poet Arnor as 'the raven feaster'. 'In helm-storm the high heart made swords sweat, crimsoned ere 15 years the claws of corbies,' Arnor wrote, his staccato lines sharp as sword thrusts, possessing a rhythm like a longboat slapping on the salt sea.

The raven as omen of war is far from restricted to Norse literature, but it is the Vikings who placed particular importance on the bird, not least its power of prophecy. In her inaugural address to the Viking Society in London in 1893, the Shetland writer and folklorist Jessie Margaret Saxby chose for her theme 'Birds of Omen'. 'I have too much respect for the corbie to dwell in detail upon its natural history,' she told the room. 'Why, when I know on the authority of a Shetland witch that the corbie can assume any form he pleases?' In her speech, Saxby revered the raven as much as the old Norse warriors who settled her lands. 'This lordly bird, swelling aloof in some inaccessible precipice, floating silently on black wings over the heads of more common creatures, dropping with stern implacable ferocity on his prey, calmly croaking of doom when the sun shines, rejoicing in a storm, haunting the footsteps of death, feasting on the dead ...'

Before their marauding expeditions, ravens were supposedly released from Viking longships to lead them towards land. This was how the adventurer Hrafn Floki (hrafn means raven in Old Norse) discovered Iceland. There are various versions of the story, but on Orkney, the Shetland version is the one I am told. It tells that Hrafn set off from the Faroe Islands, heading north in a longship with three ravens in a cage. He had stolen the birds from a raven's nest in Shetland, on a small island in the middle of a loch. When he was out at sea, Hrafn dedicated each bird in turn to the gods Odin, Thor and Baldr and let them go. Odin flew back towards the Faroes; Thor merely circled the ship; but Baldr took flight northwards, and Hrafn followed. The bird eventually led him to Iceland. Back in Shetland, however, the raven who had her chicks stolen, had not forgotten. One day she flew over the head of Hrafn's young daughter and tempted her out of the grounds of the house towards the loch. By the time Hrafn realised she was missing, he rushed out to the water and found her floating face down by the island where he had stolen the eggs, her red dress billowing about her body and the raven flying above, cackling 'corpse, corpse, corpse'.

As strict Presbyterianism gripped these islands from the eighteenth century onwards, the ravens around Orkney were viewed with as much suspicion as the old pagan Viking stories that were soon outlawed by the church. The birds were seen as having a satanic impulse – and known as 'the Devil's own bairns' – even supposedly doing more harm on a Sunday than any other day (despite the more obvious conclusion that this was when the fields were at their quietest). Still, even if the odd church-going farmer took aim at a nest, the bird's sacred position in Norse culture allowed it to breed relatively untroubled. 'The raven remains king of the cliff,' said Saxby, 'while generations of men come and perish.'

★ ★ ★

The Orkney Islands are layered with the past like sedimentary rocks. Neolithic forts and stone circles stand side-by-side with Viking farmsteads and burial mounds, old quarries, mines and wartime gun batteries. History is carved into these

Old Red Sandstone cliffs, and lies thinly veiled under the earth. Sunken ships of empires long disbanded sit barely concealed beneath the waves. Rune stones and old fragments of weaponry still wash up along the pebble beaches.

After several happy hours, I leave Fran Flett Hollinrake's cosy cottage, my stomach full of scones and buttered bannock bread and mind teeming with tales of old. She directs me north, towards the Brough of Birsay at the most northwesterly point of Mainland Orkney. This bulging outcrop of land was home to Orkney's Pictish residents (descendants of the Iron Age broch builders) who, from 600 AD built small oval houses where they lived off the land and water. From 1000 AD onwards, Birsay became a Viking stronghold. Sigurd the Stout's son, Thorfinn the Mighty, turned this outpost into the seat of power for the whole islands. The ruins of grand Norse houses, complete with saunas, are still sunken into the grass, alongside the foundations of a smithy and a Romanesque church which some claim was the Christ Kirk of the Sagas, where St Magnus was laid to rest. The brough is separated from the Point of Buckquoy on the mainland by the waters of the Brough Sounds. In Old Norse it was called Byrgisey, the Fort Island. It remains only accessible (by a causeway) for a few hours each day, when the tides draw back revealing sharp rocks like rows of glittering teeth.

I arrive in time for the late afternoon crossing, passing the final few families making their way back across the causeway, and head to the ancient Orcadian seat of power with the island to myself. Fulmar-petrels wheel about me at dizzying speed, their gurgling cackles echoing about the brough. The fulmars are in full breeding season, during which, if you strayed too close to their clifftop nests, they would quite readily regurgitate the contents of their stomach – a treacly vomit thick enough to coat a rival bird's wings so much it cannot fly. Waves pounding a constant beat and with one eye

on the encroaching waters of the causeway, I walk among the
Viking ruins, the 1,000-year-old stones set out as neatly amid
the turf as a Victorian landscape garden. Then I hear a raven,
'corp, corp, corp', rising up over a hill and disappearing
below a north-facing cliff.

I follow without hesitation, for now the instinct is rooted
deep within me. I walk over the heathland towards the point
where the cliffs are at their steepest and the wind snatches one
towards the edge, keeping my eyes ahead for occasional
glimpses of black, and tripping over numerous rabbit warrens
in the process. The raven leads me around the island
clockwise, dipping in and out of the centuries. I lose sight of
it for a while then we reconvene at an old unmanned
lighthouse: a turreted white, gold and glass structure built in
1925 that looks better suited to a Las Vegas Hotel than this
wild isle.

The raven charges on and two of the fulmars rise up to
meet it, their white wings whirring as they fly either side of
the bird. They try and knock the raven off course, weaving in
and out of its flight line, but it merely turns 45 degrees
without losing a single second of speed, coasting momentarily
on its side before tucking its broad wings back and rocketing
forwards. As it twists and turns the raven's colours shimmer
like oil floating on water, its constant calls rasping in the salt
air. The fulmars screech, clutching at nothing but the sight of
its wedge of tail feathers leaving them behind, before the
raven aims back towards the Viking settlement and again dips
out of sight. From my vantage point I can see either side of
the causeway begin to boil up in the rising tides. I follow the
bird back towards the mainland before I am cut off for good.

In the evening during high summer here, the sun rarely
sets beyond a curious twilight that suspends time in its grey
canvas. It is getting late, though I am not carrying a watch to
know exactly how so, and not yet needing to eat, or settle for

the night, I decide to head eastwards along the coast from the Point of Buckquoy. The path winds close to the clifftops before passing down towards a gentle, sloping beach known as Skiba Geo, where the Viking fishermen hauled up their boats. A turf-roofed, nineteenth-century fisherman's hut still stands at the top of the beach, solid despite the weather-battering it must receive, only one sagging wall beginning to give way to the elements. As I peer through a hole into the darkness inside, I hear the croak of a raven and look up to see one shooting by at staggering speed. I have never seen any raven fly so fast. It dives down, and yet instead of disappearing from sight, remains at eye level, parallel to the cliffs. The bird flies for a mile or so before reaching an old whalebone monument, mounted on the top of a nearby viewpoint. It banks in an arc and then flies back past me fully into the wind, slicing through it with ease. Like a slingshot, the raven then turns and rockets past me again. This time it sails beyond the whalebone and disappears, leaving me breathless at its speed.

I trudge on encumbered by human failings and a heavy rucksack. The rucksack holds a tent, and I am looking for a spot to pitch up on the clifftops for the night. The light is slowly draining from the sky, turning the ocean a milky blue. Even over the past few hours I have been walking, the weather has turned in endless directions. When the wind picks up it is enough to make my waterproof jacket bloom like a billowing sail and send me teetering towards the edge. Occasional shafts of light have forced through the clouds, illuminating patches of sea and transforming the water from dull grey to gold. The bursts of rain are so delicate and fine I do not even realise I have been soaked, until I look down at my sodden boots. There is not another soul around.

The old Orcadian dialect contains many specialist terms for these sudden changes of weather. *The Old-lore Miscellany of*

Orkney, Shetland, Caithness and Sutherland by the scholar Dr
Hugh Marwick, contains an impressive glossary. In 1929 he
compiled a dictionary of the Norn language, which is mainly
derived from Old Norse words and, by the time of his
research, had all but disappeared. The Orkneyjar website, a
modern compilation of long-lost island traditions, notes that
in the ancient language the weather was always personalised:
he rather than *it* as the necessary pronoun for describing
which particular squall was blowing in.

Marwick uncovered 'over a score' of Orcadian terms for
both wind and rainfall. For a strong breeze, he recorded: a
tirl, gurl, gussel, gushel, hushle, skolder, skuther, guster and *gouster*.
Fiercer gales were called *skreevar*, from the Old Norse *sveifla*.
For mist and rain, he attributed three distinct groups of
words. For a fine drizzle: *driv, rugg, murr, hagger, dagg, rav,
roostan, hoger, eesk* and *fiss*. A passing shower is termed by
Marwick as a *dister*, and when the sky begins to clear it is said
to be *glettan*. Reading Marwick's glossaries, one is struck by
the language we have lost. I search my mind for the English
equivalent to describe the clifftop weather and there is
nothing to compare with these ancient terms. A 'Scotch mist'
perhaps, as Marwick suggests, for the greyness slowly closing
in, swirling sea into sky and obscuring the views of Birsay in
the distance. The Orcadian poet George Mackay Brown calls
it a *'sea harr'*.

Not wishing to go too far in the enclosing fog, I pitch my
tent on a small, covered ledge sheltered from the wind. The
ledge is just big enough to accommodate the canvas, with a
patch of grass to soften my bed for the night. I cook a simple
meal of rice and, not being able to see beyond 6m (20ft) or so
in the fog, climb into my sleeping bag to read in the twilight.

When I pitch my tent, the sea is calm and lapping some 9m
(30ft) below me at the bottom of the cliffs, with sharp-edged
rocks still poking out from the shallows. Having seen the

causeway swallowed up several hours ago, I presume that the tide is already in. Yet after reading for an hour or so, until I start to doze into my book, I begin to notice the sound of waves pounding the cliffs. I listen for a time, lulled by the ocean noise, then decide to unzip my sleeping bag and climb out of the tent to check all is OK. When I look out, I discover the sea has risen up the cliffs. Every rock I had seen sticking out of the seabed has now disappeared into the waves breaking just a few feet below me. A thick mist has ghosted in with the waves and I can taste flecks of saltwater on my lip. I notice a grey seal which has also moved in with the tide to feed around the cliffs and is bobbing about in the water not far below my tent. We stare at each other through the silence. I wonder what it must make of me, and how it must see the fear in my eyes.

As fast as I can, I pack away my tent and clamber back up the cliffs to safety. I look behind me to see the seal, but it has already dived back down and been swallowed up in the mist hanging over the water. I stand there for a while, watching the water rising and wondering what would have happened, had I slipped into my sleeping bag and fallen into a deep sleep? As the water begins to lap at the same ledge where I had cooked my supper only a few hours before, I turn away. Eventually I decide to re-pitch my tent close to the whalebone monument, content it has been there for at least a couple of centuries without being washed away. I spend the night high up without any shelter from the elements, and the wind gnawing at my guy-ropes. But really, what do a few hundred years mean at the edge of ancient Orkney? What narrow places humans and their history occupy when the tides come marching in.

* * *

Mid-August, morning, and time hangs still in the constant light. The frothy clouds of meadowsweet lining the road verges and coastal heaths are slowly turning to brown, a sure sign that the dark months are coming soon. The clifftops are studded with purple Scottish Primrose, like the ones I saw in Caithness, slowly wilting from their second, and final, bloom of the year. Bumblebees drone over bird's-foot trefoil and yarrow flowers as small and intricate as pin cushions. We tread between white stars of grass of Parnassus, our boots squashing into the spongy earth. The Atlantic Ocean booms 18m (60ft) below us. Peat and salt tangs in the air.

We are aiming for a fissure in the clifftops, the length of a football pitch but only 3 or so metres across. These cliffs feel the full force of the worst storms the Atlantic can conjure, and centuries of salt spray have carved gaps into the mainland as neat and narrow as a slice of cake. The Vikings called these *ramnagills* or *ramnageos* – the *ramna* an Old Norse word for the ravens that nested there, the gill meaning a deep glen or a ravine. The cartographers who mapped these islands, pushing north in the wake of the Jacobite Rebellion, kept the same name for these ravines. One can still find them on modern maps dotted along the coastline. As with so much of Orkney, the land the Vikings claimed remains their own.

I am walking with a man called Chris Booth, a quiet, contemplative figure with a thick, grey beard and walking stick who knows more about ravens than anybody I have previously met. He and his wife, Jean, moved to Orkney in 1969 where he worked as a dentist. Ever since, he has been studying the islands' ravens, and while he refutes the word 'expert', he is an impressive authority on the birds. His interest was piqued as a youngster growing up in Cornwall in the late 1940s. These were the days when egg collecting was still a normal activity among ornithologists, and Booth admits he once climbed up to a nest to gather a clutch for himself. He remembers his raven eggs as having a light blue background and marked with deep green and brown blotches. 'We only took the eggs once, actually,' he tells me with a slight hint of shame.

Booth made it his mission to document all the raven nests on the Orkney mainland every year between 1980 and 2010. In that time, the number has doubled from 25 to 56. The majority of the nests he finds are established in the most precipitous perches, but he has also discovered nests in agricultural hoppers and inside tumbledown stone sheds; the birds rising like black smoke from the fallen chimneys of long-neglected crofts. Sometimes, at least once or twice in any year, when the farmer discovers a raven's nest on his land, Booth has returned to find it shot out. Occasionally, though, the birds are welcomed for their ability to drive away black-backed gulls and hooded crows: the young vandals of these farmsteads.

Booth has been ringing ravens for years, and has written and published several papers on their behaviour. 'The young are very docile when you ring them,' he says. 'They just lie there. I've had the parent above my head dancing and other times they disappear. Some nests, the raven will swoop down.' He discovered that an adult pair breeds for the first time at around five years old, after spending several years establishing themselves in a territory. The oldest one of his ringed birds has made it to 18 years old. Its body was recovered on the Old Man of Hoy, a 137-metre (449ft) Old Red Sandstone sea stack off the westernmost Orkney island of Hoy which, when the light catches it, resembles a giant figure lashed by the tides.

Despite the popularly held conception that ravens do not like crossing large bodies of water, Booth has discovered that the Viking myths of them being released from ships, may not be entirely apocryphal. Three of his birds have been recovered in Caithness and Sutherland having flown some 60km (37 miles) across the Pentland Firth on to the Scottish mainland. Another had flown 228km (142 miles) to Unst in the Shetland Islands, the northernmost of all the inhabited British Isles. One of the longest distances ever recorded in a raven, though not one ringed by Booth, is around 500km (310 miles). Even so, the average distance covered by his ringed birds on Orkney is a mere 12km (7 miles). They stalk their own particular swathes of land and defend them to the death.

Over the years, he has witnessed ravens do things that even his scientific mind has been unable to process. Once on a winter walk in the foothills of the Stob Binnein in the Highlands, he came across a family of five ravens playing in the snow. He sat watching the birds take it in turns to slide down a bank, before shaking themselves off and flying back to the top to have another go. I tell him I have heard of similar stories from the Welsh valleys. While out in Orkney, in a different snowstorm early one March, Booth was monitoring a raven nest. As he left the nest, the large male who presided over it suddenly appeared in the blizzard, a hulking black against the whitewashed landscape. 'It flew level with me and followed me from 100 yards away for about 1km,' he says. On the wall of his living room in a quiet cul-de-sac on the edge of Kirkwall, he has a tapestry of a raven in mid-flight, diving down off a sea cliff. 'The things they do and the way they fly is just a great joy to see.'

There is something of the raven that has imbued Booth, not least the uneasy relationship he shares with the other walkers on the cliffs. As we approach the *ramnagill* he wishes to show me, a group of four or five people appear from over the brow of a hill chatting animatedly. He curses, knowing the ravens we hoped to surprise will have already flown off, alarmed by the strange voices. Still, when we arrive at the ravine it teems with life. Fulmars veer between the narrow gaps, whose edges are coated in white waterfalls of excrement. Their shrieks echo about the crumbling sandstone barricades. Opposite us on the sheer cliff, is a pair of nesting guillemots. Next to them is a tangle of rusting barbed wire and blue polypropylene rope that the ravens have used to make a nest, although the birds themselves are not to be seen. The lack of trees on Orkney means the ravens are often reduced to building their nests from such inhospitable, synthetic materials.

The ravens, he says, are in constant battle with the fulmars for dominance of the cliff. He has recorded more than 26 instances of ravens coated in oil spat out by the fulmars. Most of those targeted are juveniles, and sometimes as many as three birds in a brood have fallen foul of the sticky mixture and been left with their wings glued shut – a likely death sentence for incapacitated birds unable to feed themselves. The fights are bitter. Once he saw a raven dislodge a fulmar from its perch by grasping its crown feathers in its beak and snatching it up into the air. On the occasions when raven nests have failed, he has seen fulmars moving on to use them for their own purposes in a matter of days.

Waves wash slowly into the *ramnagill*. Old orange buoys that have long broken free of their moorings bob about in the swill 30m (98ft) below us. I peer over the edge and look down into the dank, cool air and the gathered ocean detritus of centuries. Vertigo grips me in the stomach and the birds screech louder. George Mackay Brown, Orkney's famous poet who spent his life rooted in these communities, often reflected upon the layers of history washed up on the island shores. In the 1971 poem *Beachcomber* he finds on the shingle a rusted, salt-leather boot, half a can of Swedish spirits, a seaman's skull and a 'barrel of sodden oranges' – not unlike those buoys I see floating far below me.

Brown died in 1996 having reached the age of 74. The obituaries remembered him sitting at his kitchen table in his council house in Stromness, writing six days a week on the same Formica surface, among the scattered crumbs of his morning toast. He read widely, but the Norse sagas and old Orcadian folklore were the influence for his greatest works. He presented the islands as containing a microcosm of all human history. In his final novel, *Beside the Ocean of Time*, the protagonist daydreams of skipping forwards through the ages. Staring down into the depths of the raven cliff I feel the reverse.

In that same novel, Brown names the steamship that brings characters to the islands *The Raven*. Such a ship never existed in real life; the North of Scotland, Orkney & Shetland Steam Navigation Company instead named its fleet after Norse saints (Rognvald, Ola, Magnus and Sunniva). Interesting then that Brown chose *The Raven* as the vessel to transport his characters through the varying epochs of his work. To my mind, the raven in his book represents the constant; transcending the limits of human history.

He also wrote a poem called *Raven* – an appreciation of the conflicts that arise between man and bird. Brown puts the raven in a wicker cage aboard a ship named *Seeker*, stranded in a northwesterly gale with its crew losing hope of sighting land again. The sailors starve and taunt the raven, gorging on beer and oatcakes without giving it a morsel. Eventually when the bird is left so hungry that it screams with pain, they release it from its cage. The bird shoots up towards the sun and then veers westward, flying like a 'black arrow' desperate for food. The sailors follow, dreaming of the fertile pastures they will settle, and eventually they alight on land. None speculate on the damage done to the raven that they have followed there.

I sit with Chris Booth on a patch of grass brushing the edge of the cliffs, overlooking a sea stack known as Yesnaby Castle. The sandstone rises out of the ocean on two legs with a green tuft on its peak like an ageing punk rocker's hair. The sky is clear and the wind barely enough to stir a white cap on the expanse of sea before us. The deep, blue water fades to cerulean as it laps about the bottom of the stack. A mammoth great skua, called 'bonxies' on the islands, hulks past the stack scattering fulmars as it flies. Far closer to us, twites and tangerine-flushed wheatears flit about on the coastal fields.

We eat sandwiches and share a flask of coffee as we contemplate the broad brushstrokes of sky and sea. The former is marbled with faint outlines of cloud and so clear the moon is already in full view, though the afternoon still lies ahead. Suddenly, in the wake of the great skua, another black shape blots the sky. It is a raven flying from the direction of the ravine where we have just failed to spot them, and it lands on the coastal meadow close to us. A few moments later, its partner lands next to it. We watch as the birds hop about the earth, waddling in their ungainly piratical dance, and every now and then lurching forward and digging into the ground with their beaks. When they lift their heads up, there is none of the tension one might expect from a bird pulling an earthworm from the ground. Nor do they seem to be searching for rabbits though Booth has watched ravens staking out warrens and recovered the jawbones of juvenile rabbits from raven leftovers. Most likely, he says, as the birds teeter about, they are caching.

Ravens, as with other corvids, are notorious for hiding things. Because the birds lack a sense of smell, instead they memorise the exact locations of where they have buried their food in order to return to it later. Ravens are also master thieves, able to map out the patches of rival birds and dig up their caches. A 2003 experiment in the US by Bernd Heinrich discovered that when surrounded by competitors, ravens will wait until the attention of their rivals is distracted, before planting their own caches.

It is not only food that ravens steal and secrete. Several decades before Heinrich's experiments, the writer Truman Capote was making pioneering observations of his own pet raven, Lola. In his 1964 essay of the same name, Capote describes how he tricked Lola into showing him where she kept her treasures – after his bird stole the false teeth of an elderly guest staying in his Sicilian mountainside home. The raven, whose wings were clipped, had been given to Capote as a present by a local maid who had trapped her as a fledgling. Capote writes about his own transformation from disgust at the bird, 'an object both dreadful and pathetic', to being seduced by her intelligence.

When the teeth vanished, Capote placed his gold ring – which he had watched the raven greedily covet – on the kitchen table after lunch one afternoon and hid behind a door. The moment Lola presumed she wasn't being watched she stopped snaffling up crumbs from the table, snatched the ring and waddled out of the dining room and down a hall into the library. From there she hopped up a chair and on to the bookshelves, disappearing into a gap obscured by *The Complete Jane Austen*. Capote lists the items retrieved from the raven's cache: 'the long-lost keys to my car, a mass of paper money, old letters, my best cufflinks, rubber bands, yards of string, the first page of a short story, an American penny, a dry rose, a crystal button ...' and, of course, the purloined dentures of his house guest.

On Orkney, mainly it is golf balls that go missing. Jean Booth is a regular on the fairway and tells me that balls are often swiped off the ground and carried away. She has seen a raven come down and steal a tee shot in the past, as well as others foraging in the rough for wayward shots. Her dutiful husband, Chris, has included searching for the balls in his daily rounds of the raven nests and found a clutch hidden among the rocks on a nearby beach directly beneath one raven pitch.

I follow the trail of the stolen golf balls all the way to beyond the 18th hole of the Stromness town course. Here, amid the hulking concrete revetments of the Ness Gun Battery, where during the Second World War two six-inch-calibre guns were trained over the Hoy Sound, a pair of ravens have made their nest. It is a mammoth construction, the size of a Moses basket and woven entirely from barbed wire, cable and old bits of driftwood, with only the thinnest lining of bracken and sheep wool for comfort. Being out of the breeding season, the nest is currently empty but, sure enough, on the ground below it is a golf ball, the hard, puckered outer surface pecked and peeled away by a raven's beak to reveal its blue spongy innards.

* * *

It is 'the grimmlins' – long after the islanders have settled down for their suppers and front doors are closed for the night. When the winds sing along empty roads and the tides pull in with a heave of shingle. When the rooks and ravens tumble up and out in spiralling ziggurats heading back to their roosts, and the last light slips off the Neolithic stone circles and the green fields in which they stand, fade into black. Twilight. When a lonely walker goes from being part of the landscape to intruding upon it.

I decide to spend my last night by an old disused quarry off the main road leading back into Kirkwall. When I arrive, the birds are already here, settled in on the ledges overhanging the slate-grey water. A pair of ravens has secured two of the best-protected ledges for themselves and sit motionless with their heads already tucked under their wings. Two more ravens appear over on the far side of the quarry and drop down on to one of the ledges sending rock doves, that were previously roosting there, scattering out like sparks into the dusk.

The ravens rest for a moment, then hop between a few other ledges until the larger of the birds takes off again with a white oval shape in its beak. From my vantage point I cannot see exactly what it is, but the moment the bird takes to the air, a stream of herring gulls pour into the quarry shrieking and tearing off in pursuit. The raven rises up and then arcs about in a movement that makes the gulls follow in formation, as if connected on some giant mobile, before landing on a high metal pole that overlooks the quarry. The raven still has the object in its beak but turns away from me. When it faces me again, the object is dropped, gone. From here it flies back down to reconvene with its mate and the pair huddle up on differing ledges 3 or so metres apart for the night.

There is an Orcadian fable about raven mischief in quarries, which is told to me on another night by the island storyteller Tom Muir. An Orcadian born and bred, with a mix of Viking and Celtic blood and something more ancient again through his mother's connections to the island of Westray, Muir has devoted his life to keeping the oral tradition of storytelling intact. After dropping out of school when he was 14, Muir 'was dyslexic, but that hadn't been invented in the late 1960s and early 1970s, so they just beat the shit out of you instead'. Muir says he wanted to join the circus but ended up tagging along on an archaeological dig, which led him to discover the islands' history. His first book was re-publishing a collection from the nineteenth-century Orkney folklorist Walter Traill Dennison. Since then, he has published several anthologies of Orcadian folk stories, digging up the old tales that had long been dismissed as works of the devil.

In a deep voice that trundles and leaps like a laden barrow over cobbles – 'when I was born they shoved a Capstan Full Strength in my mouth and lit it up' – Muir tells me one such story collected from Deerness in the East Mainland, close to where he was born. 'There was a lead mine there, the story goes, and a raven's nest on the cliffs near to where a group of men were working. There was one young man who was interested in the raven, and whenever the men had a bite to eat, he always used to throw it a bit of whatever he was eating, and the bird always flew down to have a bite for itself.

'One day when he was not around, some of the other men decided they would have a laugh on the raven. They climbed up, stole its eggs and boiled them up in a saucepan before putting them back on the nest. After a time, the raven flew back, looked at its nest and could tell immediately something was wrong and who was behind it. Soon after, the raven flew down to where these men were and stole the cap off the head of the young man who always fed it, before flying away a short distance and dropping it. The young man went after his cap, but when he was close to retrieving it, the raven flew down, grabbed it and carried it a bit further. This went on until he was some distance away from the mine. By that point, the rest of the men had gone back to work. Suddenly, he heard a terrible crashing noise behind him, and the mine had collapsed, killing all the men inside – but he had been saved by the raven.

'You can tell a lot about people from their folk tales,' Muir says, to break the silence at the end of his story. 'It's the way people look at life. Moral without moralising. It sums up our identity, I suppose.'

CHAPTER EIGHT

A Night in a Raven Roost

I am fascinated by the sounds ravens make. The hidden conversations that play out when the birds gather in their flocks; the information they exchange and positions of dominance they assume. Over my journeys in pursuit of the raven I have heard snatches of these communications: single birds barking at crows and jackdaws straying into their territories, pairs calling out to alert each other of my presence in their lands and, on the sheep pastures of Caithness, juveniles calling down a mob to join them in a feed. I want to hear these conspiracies closer somehow, to seek an explanation for why they have evolved such a highly complex range of voices, and how they use them. In Ancient Rome, 65 different calls from one single raven were recorded, each possessing its own prophetic significance. To say they utter mere 'kronks' is a fabulously underwhelming description of a raven's kaleidoscopic range.

I had known about the raven roost at Newborough Forest on the Welsh island of Anglesey for some years but never managed to visit. Each summer for the past decade, I have stayed with my wife's family in a cottage in Abersoch on the mainland across the Menai Strait from Anglesey. In recent times, a pair of ravens has established a territory around the place.

I have often sat on a warm, late August afternoon watching them soar over the seascape, wondering about the giant roost 80 or so kilometres away, which I can almost see on a clear day. This is most likely where these two ravens formed their bond and set out together to find a new home for themselves. But one does not visit Newborough Forest in summer. Theirs is a winter spectacle that often takes place well away from human eyes and ears.

The ravens first started their nightly gatherings here in the 1990s and soon began to arrive in such numbers that Newborough was, at one point, considered the largest roost in the world. On some evenings, it exceeded 2,000 birds. That mantle has since passed to a roost in Idaho, on the steel pylons supporting 711km (442 miles) of power cable along the Snake River. Still, during the cold months, vast numbers arrive here from all over North Wales, England, Scotland and perhaps from even across the Irish Sea – an epic journey that calls into question their supposed reluctance to cross large bodies of water. They come to Anglesey as juveniles or lone adults attracted by all the same possibilities that prompt humans to leave their homes and travel to foreign lands: love, security and survival. When they arrive, it is a sight and sound unlike anything anywhere else on these islands.

I am standing in the forest at dusk in late November, and the ravens are coming. Even through my binoculars, I cannot see the birds from a distance, rather they suddenly barrel into view as if travelling through a portal. Larger solo birds arrive first of all, some of whom have already been patrolling the treetops since the late afternoon to secure themselves a site. They are pursued by groups between four and six strong. In between these gangs are pairs, flying so close their wings are touching. I hear the beat of primary feathers fully extended, as ravens flash over my head.

I look up over the trees into the violet sky of a cold, cloudless dusk and what Robert Burns described in his poem, *The Cotter's Saturday Night,* as 'The black'ning trains 'o craws to their repose.' From the western edge of the forest, my attention is momentarily distracted by a buzzard breaking and the clattering of noisy jackdaws. But the vortex of ravens instantly draws me back.

Their numbers increase a hundredfold, and again. The planet Venus is showing through the branches, and the forest canopy of Corsican and Montserrat pine soon disappears in the shroud of the night. The lichen, growing in deep fissures in the bark of the trees, emits a spectral glow and the ravens take on a more elemental shape.

Almost as soon as they land in the roost, the ravens are up again. They take it in turns to shoot into the air in mating displays, climbing and then plummeting down to earth like a jet plane whose engines have been shot out, before righting themselves at the final moment and repeating the whole giddy dance one more. In flight, the ravens tip and dip their wings in crude calligraphy, sometimes at such a ferocious angle it appears as if they are upside down. For all the twisting shapes they throw somehow they still fly in perfect synchronicity.

The night is well below freezing and perfectly still, the forest the perfect amphitheatre for this performance. At one point, a rifle report from a distant, unseen hunting party stills the entire roost. Crack. For a few seconds, the ravens fall quiet, and silence hangs in the icy air. Then once more the primeval cacophony rises to a crescendo.

And yet Newborough Forest, an area of 769 hectares on the south-west coast of the island, is far from ancient. The trees were only planted here between 1947 and 1965, covering a desert landscape of loose-blown sand dunes. Some 700 years ago a particularly fierce storm carried the sand so far inland that farmers were buried inside their cottages.

The new pine forest was intended to serve a dual purpose: the root network securing the sand dunes; and commercial logging providing jobs during the lean post-war period. In this it was successful, but a whole fragile ecosystem of rare insects and wildflowers was buried in the process. Some of these species, including marsh helleborines and Northern marsh orchids, still thrive further along the coast among the slacks of the remaining sand dunes at Newborough Warren, but in the forest, they have been entombed. While loitering among the trees, I dig the heel of my boot into the humus of moss and fallen pine needles. A few inches beneath the surface is golden sand.

Two species have fared particularly well in this alien treescape: the raven and the red squirrel. The population of the latter has increased, from a handful a few decades ago to around 700, although in recent years they have been suffering from the same adenovirus disease that is threatening the red squirrel in its northern outposts. Raven numbers, meanwhile, have almost quadrupled on Anglesey since 1950 to around 75 pairs. The breeding population across the whole of North Wales is currently estimated at between 700 and 750 pairs, and that is to say nothing of the juveniles swarming in over winter.

The success of the raven on Anglesey marks an intriguing realignment with history. The island is, after all, regarded as the last outpost of the Celtic Druids – for whom the raven was a sacred bird. Brân the Blessed, the giant king of Celtic mythology, translates as raven in Welsh. In life, the bird was his totem, and in death he inhabited the body of a raven. Brân was decapitated, and his head is supposedly buried in the grounds of the Tower of London, still watched over by ravens today.

Around 60 AD, the Romans launched a long-threatened assault against the Druids of Anglesey; a place they feared was entwined with black magic and evil. Tacitus, the Roman historian, records the invading army crossing the Menai Strait in their flat-bottomed boats, terrified by the sight of the Ancient Britons waiting for them on the beach. 'In the style of Furies, in robes of deathly black and with dishevelled hair, they brandished their torches; while a circle of Druids, lifting their hands to heaven and showering imprecations, struck the troops with such an awe at the extraordinary spectacle that [it was] as though their limbs were paralysed.' Still, according to Tacitus, the Roman troops soon overcame their fears. 'They bore down upon them, smote all who opposed them to the earth and wrapped them in the flames they had themselves kindled.'

★ ★ ★

It is earlier in the day before visiting the roost, and I am standing on the beach at Newborough waiting to meet an Anglesey resident called Nigel Brown. It is a beautiful morning, bathed in the sort of unfiltered sunlight one normally finds thousands of metres above sea level. Across the mainland, the Snowdon Massif range is dusted with snow. The ocean air carries a salt-tang that catches deep in my lungs.

I arrive well before our agreed meeting time and drift down from the car park to the water's edge. Here, a flock of a hundred or so pale-bellied Brent Geese has settled to feed on the green algae that cling to the rocks exposed by the low tides. As I watch the geese, as listless and sated as Tennyson's lotus-eaters, a raven flies overhead, and lands on the rocks a hundred metres away. The bird hops about before flying up, dropping something and then landing on the same spot – a process it repeats several times. A few minutes later, it flies back over my head towards the forest, and I notice a shell in its beak.

The Victorian ornithologist Thomas Nuttall was the first to record this behaviour in ravens; observing that they carry nuts and shellfish into the air and drop them on to rocks to break them up for food. I tell Nigel what I think I have witnessed, and he admits excitedly that it is something he has never seen before.

A botanist, ornithologist and until recently, curator of Bangor University's Treborth Botanic Garden, Nigel Brown has studied the Anglesey ravens from the beginning. In 2003, he produced an academic paper on the Newborough roost with university colleagues Jonathan Wright and Richard Stone. Through baiting dozens of sheep carcasses with colour-coded plastic beads, and then sifting through the ravens' excreted pellets on the forest floor, the study proved that a highly complex social structure existed within the roost. The ravens that ate the same food were also found to roost together, existing in close-knit gangs during both day and night.

The researchers also discovered lone birds recruiting others towards these food sources, through boisterous calls and elaborate flight displays. In the morning, they would fly out from the forest, in groups of around six, to the baited carcasses. The juvenile ravens, forming the majority of the roost, stray far further than established pairs, who guard their territory fiercely. These findings corroborate with an earlier study undertaken by Professor Bernd Heinrich in the forests of Maine, and were the first proof of large European juvenile roosts as 'structured information centres'.

As we skirt Llanddwyn beach and head towards the forest, Nigel Brown tells me that unlike the birds Heinrich studied in America, there is no real need for the ravens to gang up around the carcasses in and around Anglesey, as there are no other predators, apart from people, to threaten them. Perhaps, he suggests, going in mob-handed is simply the product of an evolutionary hangover; or that there are now so many birds on the island that they need to work together to get their share of the spoils.

We walk past basalt rocks piled like cushions on the sand. They are a metallic grey and studded with clumps of rock samphire growing out of the most improbable nooks. A rock pipit hops between deposits of kelp, snatching at the flies buzzing over it; its plumage is illuminated a pale green in the sunlight. A coal tit keeps watch from a stalk of marram grass waving gently in the breeze. At the end of the peninsula, where the old, stone monastery cross still faces out towards the Irish Sea, we turn into the forest.

Nigel has two simple theories as to why the roost has proved so popular. One is the location: the trees they prefer are high up on a volcanic ridge that runs through the forest, affording them a perfect view across the landscape and just a short hop to the mainland. The second is the availability of food. Anglesey is a prime overwintering area for sheep, and as a result, he says, there are plenty of carcasses lying about for ravens to feed upon.

Before the light fades, we want to search the wood for raven pellets (spat out by the birds every night), so we head off the main path and into the trees. Out of the sunlight the temperature suddenly drops, and my feet sink into the decomposing forest floor. We can already hear ravens calling out, and a few of the larger ones fly over our heads between the trees – the flap of their wings amplified by the canopy.

As we walk through the forest, I catch the scent of a distinct smell I have encountered before, but never this intensely: a close, musty odour hangs between trees spattered with bird droppings. It is a scent of decay, but not particularly unpleasant. A reminder, if one were needed, that in spite of the chits of wren and siskin between the trees, this is the land where ravens rule.

We walk past a pond where the ravens often drink and bathe. If I were to submerge my hand, Nigel tells me, leeches would latch on to the flesh in seconds – something he has experienced for himself. Fortunately, the water today is frozen, so I am not tempted to try. Nearby, I find one raven pellet a few inches long made of matted sheep wool and tiny fragments of bone. Another pellet we discover is made entirely of indigestible cereal grains. Over the years, Nigel has discovered curious fauna growing in the forest as a result of seeds dropped by the ravens: including a domestic tomato plant. Close to one of the trees, I spot a discarded mussel shell on the moss. I wonder if it was dropped by the same raven I noticed on the beach earlier that day.

Soon it is evening proper, and the birds begin to spiral in en masse. In the half-light, Nigel notices a pair on a tree branch clacking their scimitar beaks together in a display of affection. The one we presume to be the male has puffed out his plumage magnificently, so the feathers on his nape and crest stand on end. They take off after a time, to disappear further into the forest, and other birds instantly take their place.

The largest birds perch on the very top branches, their feathered beards and beaks a perfect silhouette against the stars, as they bark out their deep guttural refrain. Such is the noise and constant sound and movement of this seething chorus that I cannot focus on any one bird for long. And still more arrive, at least 400 at a guess. We stand together in silence as the ravens settle and the nightly dialogue begins.

How to describe the calls? The pig snorts, rolling logs, horse hooves on a hard road, chittering primates and popping champagne corks that come to my ears, seem far too parochial manifestations of this preternatural medley. As the night passes, we even hear snatches of raven song, a whispered ethereal sound barely audible through the chorus.

And then, deep into the evening, almost as one, the roost suddenly dies down for the night. A pair of rooks, previously drowned out by the ravens, croak out to one another in a final nightly lament and also fall quiet. The forest ebbs into a bristling stillness – made all the more so by what I know is out there in the silence. The stars burn brightly above, and we can see the Andromeda Galaxy, 2.5 million light years away from earth. Where we are standing, though, seems another world altogether.

* * *

In 1988 an academic at the University of Bern spent one winter recording all the vocalisations of raven pairs living within a 1,000km (621 mile) radius of the Swiss capital. He recorded 34 calls: 15 were individual-specific, 11 sex-specific and eight specific to the ravens in that area. An associate at the same university, Peter Enggist-Dublin, decided to build on this work by making 64,000 recordings of raven calls around Bern. Eventually, after analysing 74 individual ravens from 37 pairs, Enggist-Dublin discovered a total of 81 different raven calls.

The corvid section of the definitive *Handbook of the Birds of Europe*, published in 1994, breaks down the sounds into 14 main subsections. Stanley Cramp, the editor of the book, lists them as follows:

Pruk-call: a short barking in sequences of 3-6 evenly spaced apart. This is the raven's commonest call and almost always appears to be to a non-specific threat.

Gro-call: more uniform and softer than the pruk, used in many social contexts.

Fru-call: a softer, more whimpering version of the previous that also appears to have a social function.

Ruh-call: high pitched and at an irregular sequence. Given to establish dominance and by juveniles to attract others to feed.

Protest call: very quiet rattling call, often given as protest or low-intensity threat.

Kray-call: a loud, variable call in defence when birds are cornered.

Roa-call: a choking call advertising social behaviour.

Repeated knocking mechanical sound: given primarily by females especially in advertising display.

Trill-call: signals a high level of excitement and possibly advertising call.

Kra-call: a repetitive, husky begging.

Rrack-call: an oft-repeated warning call often given near the nest.

Squeaking call: in sexual display.

Soft cooing by pairs bonded together.

Established pairs using personal variants of innate calls.

The last one of Cramp's list was examined further by Bernd Heinrich in his own studies of raven vocalisations, which found that the birds will often mimic each other in the roost, with males and females copying the calls of their rivals. The bird's range, therefore, is endless. Reading Cramp's list, I reckon I have heard almost all of the calls he outlines in one form or another. And yet I feel no closer to understanding the true meaning of what had taken place that night. It is small wonder that science has so far failed to make greater strides into the complex life and mind of the raven.

To better explain what I heard in the roost at Newborough, I speak to a man called Chris Watson. He is a childhood friend of Nigel Brown – the pair grew up together in Sheffield on the edge of the Peak District and knew each other as young naturalists – and has a particular interest in birds. For Chris, though, it was always more about sound rather than sight. For his 12th birthday, his parents bought him a small reel-to-reel tape recorder with three-inch spools and a microphone. The machine was made by a Japanese firm called National, and 50 years later he still has it in his studio at home in Newcastle, a reminder of how the recorder has shaped his life.

'I can't even remember asking for it, but I thought it was just an astonishing gift,' Chris recalls. 'I recorded everything in the house. My mum in the kitchen. The oven. Taps. Doors. Anything. I was fascinated by the idea of trapping sound. Like time travel. One day, my parents had a new double-glazed kitchen window put in. We had a bird table in the back garden, and I remember looking out, and it was like a silent film. I ran outside, frightened all the birds off, put the seed around my recorder and then hid back inside. The birds came back, and I managed to record about five minutes of tape. I remember playing this stuff back and eavesdropping into this magical, secret world, where we can never be because our behaviour modifies it. I then became fascinated by this idea of the sculptural aspect of sound.'

After making his tape of the bird table, Chris devoted himself to sound and natural history. In his early 20s, he also became a founding member of the pioneering, electronic band Cabaret Voltaire, with whom he spent years touring and making records (his wife Maggie was a roadie for the group when they met). In the 1980s, he decided to leave the rock star life behind to pursue his wildlife-recording career.

Over the decades, Chris has worked with David Attenborough on numerous documentaries and won a Bafta for recording the 1998 series, *The Life of Birds*. Together, he says, they have travelled to 'the North Pole and South Pole and most places in between'. There have been some hairy moments, although not specifically connected to ravens. Once, after recording a den of hyenas in Kenya's Masaai Mara, Watson went to retrieve his microphones near midnight and found himself confronted by eight pairs of hungry eyes in the darkness. Fortunately, his guide came to the rescue just in time and scolded him as archly as possible in the polite Kenyan fashion, with a firm wave of his forefinger. 'I try not to think about it too much any more,' he admits.

Nowadays, he is the president of the Wildlife Sound Recording Society and was also the man behind the Radio 4 series *Tweet of the Day*, which presented a different bird and its song for a few minutes at 6 am every day for a year. In recent years, despite recording birds all over the world, Chris Watson, like his childhood friend Nigel Brown, has found himself becoming obsessed with the secret dialogues of the raven.

He first travelled to Newborough Forest a few years ago to listen to the roost. 'I was interested in the sound in that wood. The apparent conversations and communications they had. I ran a microphone cable 300m (328 yards) into the trees and left it there for a few days. I managed to record this cycle of sounds as the birds came into roost and left at daybreak.'

Unlike the night I visited Anglesey, the ravens never fell silent during Chris Watson's recordings. Because he had only left his microphones in the wood, and not been in there himself to interrupt their flow, the ravens continued a dialogue throughout the night, unaware that there were human ears listening in. 'Even at midnight and 2 am I was hearing these remarkable, quite short but conversational calls,' he says. 'It definitely sounded like a call and response. I found the quality and tone of the birds just fascinating.'

Watson is particularly interested in the way humans process the sound of the raven. He calls this the notion of temporal resolution, the speed at which information is assimilated by our brains. The raven, like other birds, processes sound roughly twice as fast as humans are capable of. 'We have to slow it down before we can even begin to understand the complexity of it,' Watson says.

Sometime after our first conversation, he sent me a few recordings of ravens calling out to one another in the roost at Newborough. I received his email one evening while writing

at my desk. When I pressed play, an unearthly sound filled
the room. He had slowed the raven-speak down to half speed,
and I could hear calls and replies from deep within the forest.
The pitch was closer to a leonine growl than something from
the avian world. Each call received an immediate response
from another raven elsewhere in the forest. That they were
birds was only confirmed by the occasional thrum of long
heavy wingbeats over the microphone. After I had listened to
the recording, Watson told me that he had never heard such
an obvious and vivid conversation taking place in any other
bird roost across the globe.

★ ★ ★

Several Novembers ago, Watson decided to take his raven
recordings into his own local woods, Northumberland's
sprawling Kielder Forest. Like Newborough, this is a man-
made landscape and the largest in Northern Europe. Tree
planting began after the government inherited the Kielder
Estate in 1932 as payment in kind for the Duke of
Northumberland's death duties. By that time, ravens had all
but disappeared from the landscape and even today are few
and far between in the forest – despite it being as perfect a
habitat for them as Anglesey.

To imagine ravens returning en masse to Kielder, Watson
proposed to the Forestry Commission that he set up an art
installation among the trees. Aided by his two sons, they
rigged up a hemispherical canopy of about 30 unseen speakers,
and invited groups in at sunset to listen to the sound of
the Newborough ravens flying into the roost. He called the
composition *Hrafn: conversations with Odin*. Named for the
Norse god's two ravens Huginn and Muninn, who would
whisper into his ears all the knowledge of the world. People
were led into the forest in groups of 15 at a time to listen to

the ravens, either lying on their backs on the forest floor, standing or sitting on tree stumps. When the recordings were played, Watson says, Kielder came alive with ghosts.

It is not long after dawn in Kielder Forest and we are walking together towards the glade where the installation took place. Watson is a tall, imposing, muscular figure in his early 60s, who has retained his wry Sheffield drawl despite his years in the North East. He prefers living in Newcastle for its proximity to the sea – as well as birds he specialises in recording underwater – and the fact that he often has the vast landscapes to himself. 'The Geordies always just spend their weekends shopping in the Metrocentre in Gateshead,' he tells me as we walk. 'He loves saying that,' chips in his son, Alex, who has joined us for the morning.

It is late November, and elsewhere in the country it's the morning after the coldest night of the year, but the forest is curiously warm. The ground is damp and several degrees above freezing. The Sitka spruce wave in a gentle westerly wind, sweeter than the harsh salt spray from the east. We are here to record the dawn chorus and whatever birds that might take part. After we have rigged up his microphones, Watson urges me to listen to the headphones plugged into a recording box, a few hundred metres away. All I can hear is the wind swirling between the trees. 'That is one of my favourite sounds,' he says. 'It's the anticipation.'

Robins wake up first, cursing chit-chit in defence of their patches. Soon we are assailed on all sides by the alarm calls of wrens, and from above by siskins and crossbills. When the morning light is still granite grey, Watson suddenly sees a large bird break from the top of a tall pine. It disappears again almost suddenly, but I am just in time to see its obvious heft. We both agree there is something of the raven about the shape, or perhaps a buzzard? Either would be a rare sighting here.

In spite of the valiant chorus from the other songbirds in the forest, silence soon punctuates our walk. Being in what should be a raven wood and yet knowing how few are resident there becomes a paradox of birdwatching, marked by what you do not see and hear, rather than what you do. In the 400 km (250 miles) of Kielder Forest there is everything for an omnivore like the raven: bogs, farmers' fields and livestock carcasses to settle on and, only a few miles away, the broad sweep of the Northumberland coastline and all the myriad possibilities of lunch that provides. Their absence here says everything about the force with which we have driven ravens away, and the fragility of their return.

Watson tells me he has seen an old, grainy, black-and-white photograph from 100 years ago that shows ravens nesting on a crag in an area that is now part of Kielder Forest. Their former presence in the landscape is confirmed by the farms, valley ridges and copses here that are still named after the raven.

In his 1889 book, *Bird Life of the Borders*, the Sunderland-born hunter–naturalist Abel Chapman describes how he saw an adult male raven shot at Hudspeth in Northumberland in 1879. He also writes about the 'insatiable insane greed' of collectors stealing raven eggs to sell at half a guinea apiece.

'There are now only a few spots remaining along the Borders where these fine birds are allowed to nest,' he wrote. 'Several of their former ancestral strongholds now only retain such names as Ravenscleugh or Ravenscrag to connect them with the memory of their ancient tenants.'

These borderlands, at one point or another in history, have been ruled by Celts, Romans, Vikings, and more modern outlaws. As we walk, we occasionally come across remnants of the old landscape: a crumbling drystone wall that now only separates banks of trees deep in the forest; a gritstone barn, steps built into the exterior wall leading up to a roof that the elements have dashed into history.

Between the thirteenth and seventeenth centuries, this area of Northumberland was the kingdom of the 'Border reivers', the clan families who pillaged north and south, and lived on heavily fortified farms, beholden to none but themselves. The word 'reive' means to rob and there is something of the raven in how the clans would spill out in groups of 15 or more to plunder neighbouring settlements for livestock, and anything else they could find, before returning at night. As much as they were feared and loathed by the ruling landowners, there was also an air of heroism attached to the Border reivers as outlaws.

Many of the Border reivers had their own ballads mythologising their life on the fringe of society, some of which were compiled by Sir Walter Scott in his 1802 book *Minstrelsy of the Scottish Border*. One of the songs in his anthology is an updated version of the English folk ballad *The Twa Corbies*. It concerns a pair of ravens perched near the body of a fallen knight discussing how best to dismember his corpse: 'Ye'll sit on his white hause-bane [neck bone] / and I'll pick out his bonny blue een / wi ae lock o his gowden hair / we'll theek our nest when it grows bare.'

The Border reivers were eventually driven out during the reign of King James VI. Ringleaders were rounded up and hung without trial. Clans were dissolved and scattered among the low countries. Often the men were conscripted to fight for the very king's forces that had destroyed their way of life. Long gone, but as with the raven, their legacy remains visible in Kielder Forest. Not least the names: Armstrong, Robson, Milburn and Charlton are all still common in the North East and can be traced to the reiver clans. Then there is the language still in use today. Blackmail, for example, which Chris Watson tells me about as we tramp through the forest.

Mail was the Tudor word for tax, and black was added on for the nefarious means through which the Border reivers would extort it.

Eventually we come to the clearing where his raven installation took place. The whole floor is thick with moss, and damp seeps in through my boots. It is a still and silent place. The only real obvious signs of animal life are the pine cones littering the ground, which have been gnawed to the nub. Most have evidently been devoured by squirrels. We pick up a few to see if any have been eaten by crossbills. The birds pluck the individual kernels with a delicate swish of their beaks, instead of the cartoonish manner of rodents.

I suddenly see my companion drop the cone that was in his hands and jolt his head up to the treetops. I hear it before I see it: the unmistakable sound of a bird now so unfamiliar in this forest; and then it banks beautifully into view. A large solitary raven wheeling about us. It slowly circumvents the clearing and then disappears once more between the trees. Kronk. Kronk. Kronk. The raven rhapsody impossible to conjure in mere words.

CHAPTER NINE

Ravens in Quarries

At which point do I realise I am becoming obsessed with ravens? It is the dreams, first of all. The ravens do not suddenly appear, but creep into the corners of my subconscious. I am dreaming about something else entirely and yet in my mind's eye I am aware of the bird, watching me.

Later, the ravens come closer. I have one recurring dream where I am chasing a leathery, flapping raven around my house; its feathers fall out, one-by-one, as I try to grasp hold of the bird. I put this down to anxiety, the equivalent of those dreams where your teeth get pulled out by the roots, and you wake up full of terror then relief that you still have molars left to bite down on your breakfast.

I have more pleasant raven dreams, too, and I write them down in my notebook to remember the details. My favourite places me on a woodland path like the ones I walked along in the New Forest, one of the earliest places I travelled in pursuit of the birds. It is night, and ravens are soaring above the tree canopy through the starlit sky. Curiously, they do not make a sound, just weave about in figures of eight. It is cold and I am watching them alone, calm and full of happiness.

Some of my real life experiences border on the fantastical. The dreams fade into the reality. Early one morning, staying alone in a motel in the Mojave Desert on my way to do a story for the *Telegraph Magazine* about a space-flight testing facility, I am shaken out of my jetlagged sleep by a tell-tale croak. I open the curtains and stand naked in the grey, dawn light watching an endless procession of ravens flying by my fourth-floor window. They come in groups of threes and fours and I count more than a hundred swirling in. At times the incoming birds bunch so close they are almost in formation, then they break apart and shoot down and up again like a roller coaster at the fair. Some bank sideways less than an arm's length from my window. I suppose they cannot see me behind the glass. It is as mesmerising as watching rays in an aquarium. My raven tank.

These ravens appear to be juveniles. They're shaggy, erratic and call out at a higher timbre than adult birds. I presume they are flying in from some communal roost in the desert similar to the one I visited in Anglesey. Later when the sun is up and I am out walking, I spot them hanging around the sprawling motel complex, perching on pylons that stretch into the distance or on the top of the flat pink roofs of assorted shopping malls. I find a raven feather sticking out of an acacia bush. Holding it up to the desert sun, the colours shimmer from brown to turquoise. I take it back to the hotel with me and wrap it carefully in tissue and put it in the bottom of my bag.

Another near dream experience occurs the morning before the day of my wedding in the Yorkshire Dales. I have gone for a walk to steady my nerves and find myself standing at the bottom of a quarry face listening to a raven call. I crane my neck but cannot see the bird, only taunting glimpses of what inevitably turn out to be jackdaws. I try following the noise, and eventually stumble across the ruins of what, from the outside, appears to once have been some sort of fort. There is nobody around, but a sign tells me this was the Hoffman Kiln, built in 1873 for the Craven Lime Company. Limestone blocks from the quarry outside were barrowed in and stacked by hand in the burning chamber, where they were mixed with coal and ignited. The lime was then shovelled out and put on to railway wagons on the rusting tracks behind me.

I walk into the burning chamber – a long, seemingly endless pitch-black tunnel interrupted every 20m (65ft) by a thin shaft of light coming in from the firebrick and stone flue halls. As I pass through the chamber, nerves on edge, I hear my footsteps rapping off the ground and the cool air raises goosebumps on my skin. The September day outside fades to black. When I get to the end and step outside I am relieved to see daylight, the last fading flowers of the year and a small stream running the colour of milk.

There is one recurring dream I have which takes me back to the flooded quarry where I watched ravens in Orkney. As with when I was there in the flesh, I hear echoes of the birds' call off the steel-grey, rippling water and watch them flying over, feathers frayed like a bin bag snared on a tree. In my dreams, however, the ravens don't just fly but occasionally tuck in their wings and dive like gannets. They split through the surface and pierce deep down into the water, bubbles streaming from their nostrils. In my mind, I follow them. When I wake up from these quarry dreams, I am desperate to see and understand matters beyond what my imagination can conjure.

* * *

The idea of ravens in quarries begins to take root in my thoughts. These forgotten blots on the land are some of the first nesting places to which ravens returned when they began their comeback across the country. The prosaic attraction of old quarries is obvious; the isolation appeals to the ravens, so too the vantage point they offer and proximity to human settlement. But there is something beyond that which interests me. I think quarries are the perfect setting to explain our symbiotic relationship with ravens and the landscape.

We cleave great chunks out of the land, dig as deep and for as long as our machines and shovels will allow, and when the ground is no longer of use, we move on; but whatever damage has been done, the ancient rhythms soon resume. When the rock chutes fall silent wildflowers will grow, as trees and ferns unfurl their roots in old spoil heaps and lorry tracks. The raven leads the charge, assuming its rightful place in this deserted land. Seeing the birds in these forgotten hives of industry makes me wonder what will happen when humans are long gone from this earth? Will ravens fly over the ruins of our great cities or, as with the Bronze Age remains of man and raven commingled together, be buried alongside us?

A few years ago, I visited Whitwick Quarry in Leicestershire with the birdwatcher and poet Matt Merritt. We parked up on the side of a busy road and climbed over a drystone wall into another world. The old granite quarry had long been flooded with cyan-coloured water, and when we scrambled up, it was secured by a padlocked gate. Still, we could see beyond it well enough. Matt pointed to one of the quarry faces where a raven nest he had been monitoring had slipped down in a recent storm. We could hear two of the birds in the air above us, their kronks echoing as loud as horn blasts against the water and rock. Then we saw a raven and a buzzard chasing after one another in quick succession, pursued by a mob of crows. On a steel rig lighting tower on the opposite side of the quarry, Matt pointed out a peregrine poised on its perch, waiting to pounce.

These quarries are part of what the writer and conservationist Richard Mabey called 'the unofficial countryside': the railway sidings, towpaths and motorway verges or 'brownfield land' in development-speak, the human wastelands where nature still incongruously flourishes. Over the course of writing this book, I have discovered ravens nesting in the most unlovely places: the old nuclear power stations of Dungeness and along the River Avon; old warehouses, silos, gun platforms and buildings gutted by fire. The ravens move into the spaces we create; occupy the scars we leave unhealed.

After my visit to Whitwick, I decide to start monitoring a quarry of my own. I choose a working quarry not far off the A1, where I know ravens are nesting. I visit every now and then, recording the movements of the ravens and whatever else I see. As I stand on the side of the slip road to the quarry, where mammoth lorries haul cargoes of limestone back and forth, I am often glared at with suspicion by the quarrymen. On one occasion when I have stepped off the path and climbed a small hill to look out over the quarry, a furious man drives over to me in a beeping gold buggy and demands to know what I am doing. 'Watching the birds?' he repeats after me, his eyes glaring beneath a high-visibility helmet in a way that makes me careful not to mention ravens. 'Well, you can't watch them here.'

But I ignore his demands and keep him at bay by sticking to the public paths, and what marvellous sights I am treated to in return. One autumn, I arrive and park up by the HGV turning circle where split rubble bags and fly-tipped office furniture are heaped by the roadside. Here I discover redwing and fieldfare flitting between the trees munching red berries. A month or so after that, I stumble across the largest flock of goldfinches I have ever seen. I count more than 50 of them – a charm – flashing silver and gold as they fly away from a large chute depositing a pile of rubble the size of a London townhouse. Deep in winter, I watch through my binoculars as two rabbits bound across a flat bed of grass-covered landfill. They take it in turns to sprint and look, sprint and look.

Perhaps I have been spending too much time in the company of birds of blood, but I cannot help but imagine them through the eyes of a predator. Another time, I watch a peregrine stoop from a treetop at the quarry edge, falling so fast and straight it looks like a zip between different worlds. The pair of ravens here are secretive and largely keep well out of sight of the road. Still, I see them every now and then, flashing over the men crushing rocks.

It is a far from peaceful site, clanking machinery and the air thick with dust all through the day. The motorway is close enough to hear a never-ending procession of traffic hurtling up and down the spine of the country. In the distance, some of England's largest surviving coal-fired power stations pump great plumes of smoke into the air. All the same, the sky is vast, and in sharp winds, the clouds scud like racing cars across it. In winter when the sun sets, it bruises over a deep magenta and orange. There is beauty in this abused and elemental landscape.

* * *

It is curious the places we feel at home. I was born and raised in central London, and the closest thing to the countryside I can remember was the urban farm at the end of our road. My extended family, however, came from the north, and on weekend visits to see grandparents in Lincolnshire and Yorkshire, I relished the freedom of those open landscapes; the feeling of being on the edge.

My paternal grandparents, Beryl and Guy Shute, lived in a village called Swainby on the western edge of the North York Moors. It is a small, traditional sort of place with a church, pub, shop and neat rows of sandstone cottages. A beck runs through the village where we used to feed the ducks. Nowadays when I visit, I sit upon a green bench by the water with a brass plaque bearing the names of my grandparents bolted on to it.

Their home was an old, white farmhouse with blue guttering on the front and stone mushrooms outside. I remember the feeling of getting out of the car after a four-hour drive up the A1 and inhaling the sweet, manure tang in the air from the farmhouse next door. It was owned by a man called Billy Bell, whose great rusting tractor my big brother and I adored.

Inside the house was a metal box of Meccano, a stuffed snarling badger and, at the foot of the bannisters, a carving of a Gibraltar ape from my great-grandfather's time running the island's airport during the Second World War. I used to rub its smooth head with my palm when I ran up and down the stairs. I still have that ape fixed at the foot of my own bannisters, and still rub its head for comfort.

The real treasures of that house, though, lay on the outside. Walking out of their front door and a breathless charge up a steep hill brought us to the ruins of Whorlton Castle. Here we climbed among the drained moat and old stone pits and imagined the time when it was full of knights, and fires roared in the fragmented hearths whose stones were blackened by centuries of smoke. Then there were the moors; with streams to dam and great swathes of heather so high, you could crawl through it unseen. Robert Louis Stevenson's novel *Kidnapped* was, and still is, my favourite book and I pretended to be David Balfour fleeing the redcoats. When it snowed, you could sled down a steep field next to the house on a binbag and fly like the wind.

I love those moors and still go there now. When my gran died, we scattered her ashes on the Sheepwash. Every so often I still camp out there for the night on a grassy ledge that juts out over a stream and is entirely hidden from the world. When I go with my wife or friends, we cook a stew, carry it up along with our tents and a bottle of wine, make a fire and sit out listening to the water rush by. The report of rifles from nearby grouse shoots sometimes wakes us up in the morning.

A long chain of memory and family leads me back to this part of the world, but in adulthood, I have become increasingly drawn to another wild expanse, this time on the other side of the Vale of York. I first discovered the Yorkshire Dales when I was a university student. It is infused with memories of my youth, in its crisp air, caramel-coloured streams and the smell of heather blossom and wildflower meadows, yet unlike the North York Moors there is no family history for me here. Somehow, though, the Dales is a landscape that puts me in my place. Satisfaction is always the word that springs to mind when I try and condense the powerful, conflicting emotions of excitement and nostalgia that being in the Dales provokes in me.

★ ★ ★

The bones of the Yorkshire Dales are the Carboniferous Limestone beds that were formed some 300 million years ago. The Craven Fault is responsible for the limestone pavements and sheer and sudden crags of the southern Dales, while the Yoredale Series in the north is interspersed with shale and sandstone, producing great plateaux of sweeping moorland. This is a landscape shaped and smoothed by glacial activity: potholes, waterfalls and caves as well as the gill, gorge and crag. When it rains water gushes from every available crevice.

In the uplands, the soil is acidic and poor: scrubby grasslands, bracken, moor and cotton grass bog. The only farming possible is livestock grazing, and it is this which has shaped so much of the modern Dales, alongside the pursuit for what lies under the earth, in particular, the rich lead deposits that vein through the valleys.

I can be here and watch history unravel. I look at the topography of the Dales, so different as I travel between each valley, and picture the bed of a tropical ocean millions of years ago. At natural landmarks, like the overhanging limestone lip of Kilnsey Crag, I imagine the transhumance of nomadic herders driving their flocks down to winter pasture and to trade at sheep fairs. Passing by train over the Ribblesdale Viaduct, I picture the poor nineteenth-century navvies who built it. Wandering between the old lead mines and limestone quarries that still scar the top of crags, I think of those who lived here in tent cities like men on a gold rush, but with no promise of riches at the end. Walking over the now bare hills, necklaced with centuries of drystone walls, I imagine the great forests that once grew here.

Lead was being mined in the Dales as far back as the Roman era, and during the eighteenth and nineteenth centuries the industry boomed. In Britain's lead heyday between 1821 and 1861, more than 20,000 tonnes were extracted from Grassington Moor alone. It was bleak, dirty work. Tunnels were only a metre tall and prone to caving in. Men, women and children all spent long days on the open moors, cracking open the rocks and extracting the lead ore by hand. By the end of the nineteenth century, the industry had collapsed, but the hills are still covered with the debris left behind – the flues, smelts and mineshafts. Close to the peaks of many hills, you can still stumble across the skeletons of old lime kilns (like the Hoffman Kiln) and old limestone quarries and even some that are still in use today.

I love the Old Norse names of villages like Muker, Redmire and Hardraw and pore over Ordnance Survey maps to discover more. I have other maps, too, ones of my own creation. I have spent weeks walking and cycling across the 1,400km² of the Yorkshire Dales. I have swum naked in the River Wharfe and bathed in a wild waterfall in the furthest reaches of Wensleydale. I have been burnt crimson in the sun and once, while cycling over a particularly exposed bit of Coverdale, was soaked by a downpour so terrible that I subsequently caught the worst bout of flu I have ever endured. I have slept out wild many times. I know of one spot, in particular, to peg out a tent, listen to the hooting owls and watch the moon glint off the silver, snaking riverbed below. From this land, I derive pure physical joy.

So, in my mid-20s, when I became a reporter for the *Yorkshire Post* with a beat that covered the Dales from Blubberhouses Moor in the south to Richmond Castle in the north, I thought I had the best post in journalism in the world. I wrote about sheep rustlers and planning applications, road accidents and rare orchids, public toilets and family histories, agricultural fairs and rural poverty, parish council meetings and nature. When travelling between jobs, my phone would soon run out of reception with no signal for miles to allow the office to reach me. I could vanish into stories.

★ ★ ★

One birthday, a year or so after my appointment, I received my first ever pair of proper binoculars. I was in the Wensleydale village of Askrigg and decided to take them for their inaugural outing to where the narrow main street begins to arc up and over the tops into Swaledale. Here, opposite a tree teeming with chaffinch and sparrows, I raised the lenses to my eyes and first felt that thrill of delving into another secret world.

The Dales was where I really began to fall in love with watching birds. I marvelled at curlew, lapwing, chitting snipe and oystercatchers and, one early spring morning when Wharfedale was covered with snow, spotted my first ever dipper, hopping up a riverbed towards the limestone crags of Malham Cove. So too, my first sighting of grey wagtails along the Ure and Swale.

I remember these first sightings as vividly as the ravens that stalk my dreams. People have told me they have seen ravens at key moments in their lives; births, deaths and in moments of deepest grief. The first raven I ever saw in the Dales was outside the church of St Matthew's in the tiny village of Stalling Busk. The village overlooks Semer Water, a kidney-shaped lake, which on clear days reflects the shimmering peaks of Raydale that surround it on all sides. My wife and I were taking in this view when we heard the telltale kronks and watched as a powerful raven flew into our eyeline and rocketed down the valley with its wings tucked behind it.

We later decided to get married in that church and a few months after our wedding returned for another visit. As we stood outside waiting for the service to begin, we watched another raven appear through crepuscular rays in the clouds, this time moving with long, languid flaps of its wings. What had previously been an empty sky suddenly filled with birds. Jackdaws that had gathered in a nearby copse shot into the air in pursuit, cackling and clattering as they urged each other upwards. A buzzard also appeared out of nowhere to join the fray. The raven flew between them, twisting and corkscrewing, as one by one the jackdaws tried to jab it down from the sky, making vicious lunges that the larger corvid simply shrugged off. The raven flew over us, two or perhaps three times before uttering one last kronk and returning to where it had come from. We watched open-mouthed in amazement. That was the moment I came to realise one of the great joys of the raven that I had never fully appreciated: this supposed bird of death, in fact, animates a landscape, brings it to life.

★ ★ ★

Like me, Doug Simpson was an outsider to the Dales and somebody who discovered the joy of birdwatching late. He was in his late 30s and a keen angler when he came to realise he was taking more interest in the great crested grebes nesting on the island of the lake he fished near Leeds than he was in the fish themselves. Back then local teenagers would try to get across to the lake to steal the grebe eggs, and Doug took it upon himself stop them, on one occasion chasing a gang of youths causing them to throw away the tin in which they had stashed the grebe eggs. Eventually, he sold his fishing tackle, bought a pair of binoculars and a scope, and ended up as the protection officer for the Leeds Birdwatching Club.

He first became involved monitoring persecuted birds in the Yorkshire Dales in around 1979. Doug speaks quietly and with great precision and admits that on this occasion he cannot remember the year with any certainty. He has never left.

Back then peregrine falcons had just started to return after the boom in the use of organochlorine pesticides, like DDT, following the Second World War. They had cruelly devastated populations by thinning the walls of their eggs, rendering entire clutches unviable. The impact was so severe that the bird's demise was almost total. When he assumed his post, Doug had never seen a peregrine before, nor for that matter, a raven, and was invited by the Yorkshire Naturalists' Union to help monitor potential sites where the birds could have returned. He can still recall the first time he spotted a peregrine on Blue Scar crag near Arncliffe. 'Zoom zoom,' he says with a grin.

As well as the threat of pesticides, Doug soon realised the ongoing danger from gamekeepers on the large grouse shooting estates that stretch across the Dales. 'When nests failed you couldn't be sure if it was a natural occurrence or somebody had been there,' he remembers. 'That was the fear in those days. You would visit a place, and everything looked all right, then go next time, and there was no sign of the young.'

Doug's main cause célèbre in Yorkshire has been the return of the red kite. He led the project reintroducing the birds to the Harewood House Estate near Leeds in 1999. The scheme has proven such a success that Doug was appointed MBE. While he continues to be heavily involved with the monitoring of the county's burgeoning red kite population, he also still focuses on peregrines and the king of the corvids that he refers to as an 'honorary raptor' in the Dales.

Ravens followed the peregrines back to the Dales. A lone pair nested here for a spell in the mid 1980s and then, after a few years, suddenly disappeared. Their numbers have slowly increased since. In 2008 there were nine occupied breeding sites, of which six were successful. In 2015, 10 breeding sites were occupied, in nine of which 27 young were raised.

After reading in the *Yorkshire Post* about his raven monitoring work in late 2016, I emailed Doug asking if I could accompany him on his winter recording of raven nests. He agreed, but only on the condition that I kept the various locations secret. The ravens, he explained, still had many enemies in the Dales.

★ ★ ★

I meet Doug at his home in Harrogate on a murky December morning, and we drive out together into the Dales. He is wearing what I later discover to be his customary bird-watching outfit, a thick brown and olive woollen coat and trousers. Ravens are the earliest nesters of any bird but still wait until the worst of winter is over before laying their eggs, in all likelihood around February. At this stage of the season we are only planning to visit a few sites where they may be rebuilding their nests. If undisturbed, pairs will return year after year to the same nest, fortifying it with new materials ahead of each annual brood.

We also want to see how the ravens are interacting with the peregrines, with whom they often share the same nesting sites. Both species favour the steep, wild, limestone crags and old quarry faces of the Dales to rear their young. While the raven builds its impressive towering nests, the peregrine normally relies on tiny depressions in the cliff face known as scrapes, though Doug tells me he has previously recorded peregrines occupying raven nests.

In a 1962 study of the breeding densities of ravens and peregrines, the ecologist Derek Ratcliffe noted a 'proximity tolerance' limit between adjacent nesting pairs. Famously, this tolerance is often pushed to the point of downright hostility. I have heard and read stories of vicious dogfights between ravens and peregrines. One bed and breakfast owner in the New Forest told me that she and her husband would sit out on summer evenings watching the birds compete for a pylon, going at it hammer and tongs, flipping over one another to assume aerial domination. In his journal entry for 16 March 1924, the ornithologist Ernest Blezard describes another such encounter in the Scottish Borders, where a female peregrine stooped so furiously upon a pair of ravens that they were forced to take refuge under a boulder.

After stopping off at a working quarry where ravens have been nesting for the past three years, we visit an old,

long-unoccupied limestone quarry where Doug is one of the
very few people to possess a key. As we open the gate and
walk in, he tells me that in all likelihood he is the first person
to visit since the previous nesting season.

The sense of space and isolation here is total. A moonscape
overlooked by the quarry face rising hundreds of feet above
our heads. Wild strawberries have sprouted all over the
gravel-bed floor, which is the size of several football pitches,
and young alder and older gnarled hawthorn are slowly
reforesting the earth. The ravens, Doug explains as we walk,
nest on one side of the quarry face and the peregrines on the
other. All we can hear are jackdaws caterwauling and darting
about the quarry like a Greek chorus announcing our arrival,
as we trudge over the gravel to set up a scope.

Doug soon finds the ravens' nest, expertly camouflaged
in spite of its size, on the top of a triangular stone plinth set
into the cliff. It appears as though fresh sticks have been
added to a construction that is already the size and shape of
a beach ball, but there is no sign of the pair. We watch for a
bit and then decide to move on to get a slightly better view.
It is then that the piercing alarm call of the falcon strikes
up. 'Peregrine,' Doug whispers, and we both freeze on
the spot.

It is the male, known as the tiercel, that has taken off and
seems to be flying in an awkward, laboured manner for such
a graceful bird. The jackdaws rise to meet him in black
wisps and he twists to avoid their parries. We realise the
tiercel is carrying something in his talons, a vole or rabbit
perhaps? Certainly, something large enough to weigh him
down. The female peregrine, larger and with more brown
in her plumage, is suddenly also up, slicing effortlessly
through the cloud of jackdaws as though she belongs
somehow to a different sky. She performs one graceful
pirouette above our heads and then lands back down again

on a grassy ledge close to their scrape. As she does so, a mass of white feathers explodes around her. We realise now what the fresh kill was. The peregrines will be eating pigeon for lunch.

That first day exploring the Dales together we only see one raven. Close to sunset with the winter shadows already lengthening, we are tramping over bracken high over Arncliffe, named the Eagle Cliff in Norse, when I hear the raven call. The bird is already over us, coasting on thermals high up like a buzzard, with its feathered fingers extended out like saw blades. We count four different tones of calls from this single bird alone: pruks, kronks, croaks and grunts. We are breathless when the raven is there, and despite being strangers that morning when it is gone, we clap each other on the shoulders. I look at Doug and he is grinning from ear to ear. 'Seeing a raven always brings a smile to my face,' he says. 'It's one of those wonders of nature.'

★ ★ ★

Another day a few months later and we are back in the abandoned quarry. It is early spring, and a thick bank of mucky, grey cloud has rolled in, squatting like a toad over the Dales. During the course of our drive that morning, we notice the lowland fields have flooded and the rain continues to fall in squalls. It is early March, and the ravens will, in all certainty, be on their nests by now.

They protect their clutches against such cold weather by lining their nests with sheep's wool and laying the eggs deep within them. The female normally incubates the eggs for about 21 days before they hatch. Ravens can lay four to six eggs – quite large clutches relative to most birds – and fledge about three young. The family then stays together until early summer, the young ravens learning to fly off the quarry edges, before they disperse to join flocks of other juvenile ravens. When the young have left, the adult pair will begin securing their territory before the next breeding season.

When we pull up at the quarry we are surprised to see a For Sale sign outside. The site has been closed for 10 years after decades of intensive quarrying, but is now being touted for a 'wide range of leisure or commercial uses'. Were it to be sold under such loose guidelines then the ecosystem that has built up in that time, the breeding ravens and peregrines, would all be under threat. An entire nature reserve once more juddering apart under jackhammer and pick.

As we begin walking into the quarry the whole place today seems muffled by cloud. Even the jackdaws are uncharacteristically silent.

We climb up the same steep slope to the barbed wire barrier, beyond which there are piles of rocks that have collapsed from the cliffs above us. It is only when we set up the scope that I notice Doug has cut his thumb on the barbed wire. Drops of blood roll down his nail and drip on to the floor.

A metre or so in front us is the body of a rabbit lying on its back. It is disembowelled, with its ribcage split, and opened up like a book. There are deep slashes through the fur and down to the tendons on its haunches, which suggests that ravens rather than peregrines have been at it. That, and the fact that unlike ravens, the falcons prefer their food as fresh as possible. I am excited to see the rabbit, as it seems confirmation that the ravens are nearby, but I notice Doug is giving it a suspicious look.

We have to be careful not to disturb the birds by straying too close, but the thick cloud makes it almost impossible to see. We set up the scope on the nest we had previously recorded on the stone plinth, and after several minutes of taking it in turns to squint through the eyepiece, we think we can see a silhouette in the bottom left of the nest. Because ravens are so secretive when they are upon their eggs, it is hard to tell whether or not the bird is actually there or if it is just a trick of the light. We do not want to go any closer, so instead, we sit and wait, taking it in turns to keep an eye on the nest.

The wind has picked up cold enough to prick tears in my eyes. The jackdaws have been stirred up now too, dipping about us in and out of the fog, yaffling their displeasure at our intrusion. Or is it us? Suddenly we hear the shriek of the peregrines on the opposing cliff edge and both of the birds are above us. We have the best sight of the female who rockets down to our left, and then rises up and out of the quarry in ascending circles of ever-increasing size and speed. Before she has risen above our heads, we can see the chestnut plumage on her top, which flashes into silver breast feathers and yellow talons the colour of egg yolk.

Doug looks back through the scope towards the raven and urges me to see for myself. This time I can see the raven's head bobbing up and down on the nest as she warms her eggs. We are delighted that, for at least another year, ravens will be able to call this quarry home. Not wishing to disturb the bird any further, we walk back down towards the car. The rain has ceased and cloud thinned a little. Among the trees, I hear dunnock and goldfinch striking up to celebrate the short respite from the gloom. When we are at the car, I look back towards the nest through my binoculars. It is only a fleeting glimpse, but I see a large black bird – far bigger than the jackdaws – slip over the top of the quarry and glide down towards the direction of the nest. Behind it a rainbow has formed.

Doug admits to me a few weeks later, when we speak on the phone, that he was worried about the body of the rabbit we discovered and has been thinking about it ever since. I had presumed it was a corpse, which the ravens had picked up and carried back to their nest, but Doug raises another possibility. There is a public access footpath over the top of the quarry and, he says, it would have been easy for somebody to bait the rabbit with poison and sling it over.

This seems overly suspicious, paranoid even, but then Doug has witnessed enough in his 35 years monitoring birds in the Dales to make him so. For the past 20 years, North Yorkshire has been the worst county in England for bird of prey persecution. Near to some grouse moors, those monitoring birds have had their tyres slashed, while not long before I contacted Doug, seven red kites had been discovered (shot or poisoned) in the space of two months. Like scavenging red kites, ravens can be tricked by poisoned carcasses. At one raven nest in the Dales a few years ago, Doug recovered the bodies of three fledgling ravens. They were at that point so badly decomposed, that the laboratory technicians who were supposed to test the bodies for toxins, merely disposed of them instead.

Incidents of poisoning remain an all too familiar occurrence here. One red kite killed in 2016 was found to have ingested eight different types of poison, three of them banned substances. Shooting is similarly common. That same week we visited the nest, Doug had been alerted to the body of a red kite discovered in Nidderdale, the first that year. Scans revealed it had been peppered with shotgun pellets. So frustrated were some in the Nidderdale community with the ongoing persecution of raptors, that a local businessman offered up a £1,000 reward, prompting contributions from others to raise the total to £4,000. Doug is resolute over where he feels the blame lies, but admits in the closed world of gamekeepers it is unlikely that anybody would speak up.

'For the most part, we're talking grouse shooting areas,' he says. 'It's as though anything that isn't a grouse or is a potential threat to one is deemed the enemy. So long as gamekeepers remain in that form of gainful employment they are always going to maintain that view. It's a subject that keeps coming up in conversation. How can somebody just stand there with a gun and deliberately destroy a beautiful creature like that? But we are talking here about people quite accustomed to killing. All the time they are trapping and snaring. They either have it to begin with or develop the particular mindset which completely disregards any concern for the creatures involved.'

I want to speak to a gamekeeper in the area and ask around a few old contacts from my time working in the Dales. But I am told, in no uncertain terms, that raptor persecution is a subject nobody who works on the big shooting estates will be willing to discuss with me – on the record at least. Eventually,

I try a man called Brian Redhead, the former gamekeeper of Lord Lowther's estate, which spans 30,000 acres of the Lowther and Eden Valleys to the west of the Yorkshire Dales. Now 74, Brian retired from his gamekeeping duties to become one of the most prominent raven breeders in the country. Four of the ravens that appeared in the Harry Potter films came from Redhead's aviaries.

He is a devout wildlife enthusiast and lover of all birds. The morning we speak he has been out in the valleys watching black grouse lekking, and he is also one of the few people I have met, aside from the sheep crofters in Caithness, who will openly admit to having shot a raven. He did so back in 1962 when he first started out gamekeeping in Argyll on the west coast of Scotland, before the birds were protected by law.

'I was 19 years old,' he says, 'and back then you did what you were told to do by a head keeper. I was standing in a rowing boat and he pointed out a raven on a clifftop. I shot it off the top of a cliff 200 yards away with a 2.2-calibre rifle. I felt pretty good at the time when I saw the cloud of feathers. It was a hell of a shot. I suppose I did regret it in a certain sort of way but that was what you did. It was part and parcel of the job.'

By his 30s, Brian had moved to Cumbria and began working on the Lowther Estate, which at the time was owned by the seventh Earl of Lonsdale and with a prominent grouse shoot. Brian was charged with running the Lowther Park shoot, which took place in the surrounds of the old Lowther Castle. In the grounds, he spotted a pair of ravens nesting on a 36-metre (120ft) Scots pine. 'On a low ground shoot, they never cause problems,' he says. 'Up on the moorland they would definitely predate grouse nests. Back then the principle thing was anything that was considered vermin had to be got rid of.'

This time Brian was in charge and decided to keep the ravens where they were. Different eras have different views of conservation, as the old memoirs of Victorian and Edwardian lovers of wildlife attest to. Many of the authors of these books boast of stealing rare eggs, shooting birds and having them stuffed or dissected to further their understanding of various species. In Brian's belief, there is no dichotomy between shooting and loving birds. He took his first bird egg at the age of eight and fired his first air rifle at 14. 'You don't have to be a mass murderer to like shooting,' he says. 'It's not strictly true that gamekeepers aren't supposed to like ravens. They may be a nuisance at times, but it's not like you're going around annihilating every raven you see.'

As well as gamekeeping, he has kept and bred numerous wild animals: Scottish wild cats, pine martins, choughs and eagle and snowy owls. His first raven arrived about 30 years ago and was a 21-year-old female given to him by a local wildlife park where she had become surplus to requirements. Brian kept her in a large aviary in his back garden and soon became intrigued. The more he studied the birds, the more he became convinced that they were not really birds at all. When he racks his brain for a comparison he eventually settles on gorillas.

'When a gorilla looks at you it eyes you up and down just like another human would. It's almost like it's trying to read your mind. A raven has that same potential to understand what you're going to be doing. I don't think you get that kind of relationship with any other type of bird. You can see their intelligence.'

It was because of the ravens' intellect that he decided to start breeding them. He deliberately chose not to hand-rear a newly fledged raven because of the strong emotional attachment between human and bird that he knew would ensue. Instead, he took satisfaction from getting the ravens to breed – and then sold the chicks on for £400 a piece. He estimates that over the years he has probably sold dozens. He also never gave his ravens a name – even the first female he owned. If you give them names, he says, that turns them into pets and he had far too much respect for his ravens to reduce them to that.

In June 2009 after he had retired from working on the Lowther estate, Brian went out to visit the aviaries, which he kept in a field about a mile or so from his house. He discovered that three of the four aviaries – each about 7m (24ft long) – had been damaged by vandals overnight. A door to one had been kicked off its hinges and a padlock broken, as well as another side panel smashed. The culprits had also cut through two-inch wire mesh on the roof of the aviaries and peeled it back. In total, he lost five pairs in the attack, one of the female birds dying of what he believes was a stress-related condition the following week. Despite offering a reward for information in the local paper, nobody ever came forward. Nowadays Brian keeps corn bunting, cirl bunting, hawfinch, redwing and numerous other species, but never again a raven.

'I miss them a lot now,' the old gamekeeper admits. 'It's just something you can't put your finger on. Whenever I see a raven flying in the wild, I always wish I still had my ravens. They're always with me, and they are larger than life. I love all birds, but the raven is just like …' he trails off trying to seek the right words. 'It's difficult for me to say,' he eventually says, with emotion pricking his voice.

★ ★ ★

It is a fine spring morning, Doug and I are stood in a farm in Upper Wharfedale watching newborn lambs gambol in the fields while two ravens soar overhead. A radio blares in a barn behind us and to our immediate right is the corpse of a young calf that has died overnight. I look at its crumpled, lifeless form and see the imposing limestone crag in front of us reflected in its glassy, brown eye, which, despite the presence of the ravens, remains intact.

The juxtaposition of lambs and ravens instantly reminds me of the crofters in Caithness I met the previous spring, who complained of being under siege by juvenile mobs of the birds, but here, Doug tells me, they happily coexist. A pair of ravens have secured the crag as their own territory and nest, leaving the livestock well alone. Lapwing and curlew nest in the fields in between, and as we watch, we see both species of bird bursting up from the pastures. Skylark, too, lend their voices to this joyful chorus. The sun is shining bright enough to pick out the intricate faults in the limestone face; worn and cracked in patches and elsewhere smooth and gleaming like a weathered piece of bone. The scene is one of harmony in the uplands, of man, beast and bird all thriving together.

There are two raven nests on this crag, both imposing almost cylindrical structures that have been a home to the birds for many years. We are looking at the one on the far left of the crag for raven activity, but instead notice a jackdaw rifling about on the stone ledge and flying off with a twig. This means the ravens have, for whatever reason, vacated it this year. We scan along the crag, probably 30m (98ft) from the top to bottom with trunks of whitebeam and hornbeam jutting out at improbable angles, and focus instead on the second raven nest built deep inside the gloomy recesses of an overhang.

As I am watching, a raven – the male we presume – slips out from the nest and takes off into the sky. It barks loud commands, which echo across the crag, and as it flies accelerates at will. It reminds me of cyclists in the Tour de France when, during mountain descents, they drop back their shoulders and pull their arms in to become at one with the bike.

Once the raven has flown in a long lazy arc around the crag (in a supremely confident manner), it settles on the bare, startled branches of an old hornbeam at the top of the cliff face, which looks like it was struck by a bolt of lightning. The raven begins to preen itself with its beak, digging deep down into the shafts of its feathers. The birds tend to moult each year after the breeding season, and as it burrows and pecks it dislodges one large, primary feather. I follow the feather through my binoculars as it slowly drifts down to the base of the crag and settles in a tuft of grass at the bottom. Were it not for the breeding birds all about us, I would be tempted to climb over the drystone walls, that separate the fields from the crag, to try and retrieve it.

A potent symbolism is attached to the raven's feather, whose barbs are deemed interwoven with prophetic fallacy. In *The Tempest*, Shakespeare has Caliban curse Prospero and Miranda:

As wicked dew as e'er my mother brushed
With raven's feather from unwholesome fen
Drop on you both!

As with so much raven mythology, there is a flipside to this. In other legends, the raven feather is regarded as a symbol of great fortune. Watching the feather float down the crag face reminds me of a Victorian parable I once read called *The Raven's Feather*. In it, a starving orphan is praying, on Blackfriars Bridge, that he will not die of hunger when a raven's feather drifts to his feet. The boy puts the feather in his hat as a keepsake and soon afterwards is adopted by a carpet maker called Mr Raven. The story

quotes some of the numerous Bible scriptures that make
mention of the raven including: 'who provideth for the
raven his food when the young ones cry unto God?' from
the Book of Job, and in Luke XII, 'consider the ravens; for
they neither sow nor reap; which neither have storehouse
nor barn; and God feedeth them'. As with the old Viking
legends of warriors releasing the birds at sea, Noah sent a
raven from the ark to find land after his boat was grounded
on Mount Ararat.

We follow the sun around the Dales that day. When it
moves, so do we, driving in Doug's Skoda between valleys
rinsed by light. At lunchtime, we find ourselves watching a
deep ravine where the ravens have nested before, munching
our sandwiches in the shade of a hawthorn, and looking
for movement in a nest between two juniper bushes far
below.

The ground is studded with bright yellow clumps of
celandine, and the air carries the smell of the fresh river water
of Cowside Beck, snaking along the valley bottom several
hundred feet below. The sun is warm on our faces and we sit
on foam mats to stop the damp leeching up from the long
grass into our trousers.

Suddenly a kestrel rises up hunting. We are above the bird,
which hangs still in the air pumping its wings and adjusting
its splayed tail feathers through tiny twitches. The kestrel
dives several times, never aware of us watching its chestnut
back and pale grey head. Doug explains it is most likely
hunting for voles among the scree slopes. Because of its
specialist ultraviolet vision, it can trace its prey from the
urine streaks left behind on the ground – in a similar manner
to a police helicopter using heat-seeking sensors to track
criminals on the run. Over the next half hour, the kestrel
floats back and forth down the valley. Elsewhere, out of sight,
we hear a succession of three loud cheeps. It is a ring ouzel,

the blackbird of the uplands; Doug's first of the year and mine in a lifetime.

We talk a lot about a 'sense of place'. As a gift after one of our days out together, Doug presented me with an old Ordnance Survey map of Stalling Busk (where I got married and saw my first raven in the Dales). He has never been abroad in his life, preferring instead to explore the constant changes within his own immediate vicinity. Aside from the Dales, he has another even more local patch, a scrubby area of heathland a short walk from his Harrogate home, which he is attempting to get recognised as a Site of Specific Scientific Interest (SSSI). It is an overlooked part of the countryside; in the distance are the giant golf balls of RAF Menwith Hill, a US military spying station. There are also seven wind turbines and the old barrack blocks of a Leeds University building once used to study cosmic rays hitting the Earth's atmosphere. This patch of fringeland – the sort gobbled up by local authorities across the country for cheap housing development – sustains rush pasture, purple moor-grass and the odd tangled hawthorn twisted by the elements. Currently, skylarks have taken it over. He has also counted golden plover, snipe and hen harriers shooting over the scrub. No ravens yet, but he thinks their slow, cautious spread across and out from the Dales will continue.

* * *

Our final nest to check, before the sun sets, is a distant and wild part of the National Park. It was the first place in the Yorkshire Dales to record the return of the ravens in the 1980s and somewhere I had never previously been. We drive up a steep, narrow track as far as we can go and park a mile or so from the raven nest. As soon as we step out of the car we hear a cacophony of noise. There are at least two ravens

calling out loudly from the direction of the crag we're
heading for and jackdaws gabbling, too. We shoulder the
scope and our rucksacks and strike out quickly.

As we walk, I see a blue pick-up driving across the hill
below where the ravens nest. The car disappears from view
as the path drops down between some drystone walls. We
see it again, this time it is parked up and there are three
people and two dogs walking back across the field under the
raven nest. In the sinking sunlight, there is a glint of steel
and when we look through our binoculars it is clear they are
holding a gun.

The ravens fly up from the nest. One of the bird's tail
feathers have previously been slashed in half, trimming down
its diamond shape to a far sharper angle. Still, the raven
appears unencumbered by flight as it circles around the crag
calling out in short grunts to its partner. We crouch behind a
drystone wall with our binoculars up and hold our breath at
the proximity of the birds to the shooters. Fortunately, the
entire crag is ablaze in sunlight warm enough for the clutch
of unhatched eggs to incubate, even if the adults have
temporarily left.

The shooting party, though, appears oblivious to the ravens
and seems far more interested in the rabbits. The dogs rootle
low and close to the turf in the hope of flushing them out of
their warrens. While the gunman keeps the weapon hooked
over the crook of his arm we never hear a shot. After waiting
and watching from a safe distance until the sun begins to set
behind the crag, we take our leave in the hope that the ravens
will return to the nest before the evening is upon us. I look
back several times as we walk back to the car and still see the
shadow of the birds wheeling above the crag, they appear
both an integral part of the landscape and something fragile
within it.

Doug emails a few weeks later to tell me he has discovered three young ravens in the nest. They are already well grown and looking healthy. The adults disturbed by the shooters that day had returned to care for their brood. It is a relief to read. For even in the wildest Dales, theirs is a hair-trigger existence.

CHAPTER TEN

Hunting Ravens

It is a foul day, and it occurs to me that they often are in this book. Dawn never bothered to shake off the grey, and by mid-morning the rain is from a bad Hollywood thriller – a trench coat-soaking downpour so heavy it could turn an umbrella inside out. I am sitting in a parked car without the wipers on, watching fat drops smack and then congeal on the windscreen before sliding off again in an endless carousel. Outside, a small procession of mourners makes its way over an iron bridge and into Wakefield's Walton Hall, where a funeral party is gathering.

The car park is slowly flooding. Men in suits slosh through puddles holding coats over their wives, in an attempt to keep their hair dry. My car is steaming up with breath. It is a Thursday. I try not to imagine getting buried on a day like this.

On a better morning in June 1865, another funeral took place at Walton Hall, an eighteenth-century Palladian mansion, built on an island in the middle of a lake, a few miles from Wakefield. The dead man was Charles Waterton, the 27th lord of the estate, known among all and sundry as the Squire of Walton Hall. The *Illustrated London News* described him in its obituary at the time as 'that most genial and enthusiastic of all field naturalists', who had died at the age of 82.

A flotilla of boats ferried Waterton's body from Walton Hall to his grave. According to the diary of Waterton's friend Norman Moore, the Bishop of Beverley, dressed in his mitre and robes, rode in the flagship. He was accompanied by four purple-cloaked canons and 13 priests who sang the Requiem Mass as they cut a path through the water. Behind them towed a coal barge-cum-makeshift floating bier bearing Waterton's body. Behind that was Waterton's own boat, *Percy*, empty of passengers and draped in black. Three other boats carrying mourners brought up the rear, while hundreds of others followed on foot from the lakeside.

When the boats reached the shore, Waterton's coffin was buried under an old granite cross which he had built between two old oaks the previous year in preparation for his death. At its base was a marble slab and a plaque bearing the epitaph: 'Pray for the soul of Charles Waterton whose weary bones lie close by this cross.'

Waterton travelled widely as a young man in the jungles of Guyana, making his name in nineteenth-century British society as a gentleman explorer and conservationist. He managed to cheat death countless times during his travels and returned – wracked with dengue fever and malaria – to his inherited 300 acres, where he established Britain's first protected nature reserve. While the Industrial Revolution boomed in coal country all around him, poisoning rivers, digging mines and felling woodland in the name of commerce, Waterton erected a vast three-mile long and 4-metre (16ft) high wall around his estate. It took four years to build at a cost of £9,000 (£2.5m in today's money) and was completed in 1826. Everything inside of it he devoted to the preservation of animals.

His years exploring foreign wilds had taught Waterton to eschew any of the comforts that typified the landed gentry of his time. He slept each night on a wooden floor wrapped in his

Italian cloak in front of the fire, with a block of smoothed oak
as a pillow. He ate little, drank nothing, wrote prolifically, and
banned all smoking from his house. He also took a macabre
delight in serving his guests carrion crow pie and pretending
the meat was from the fat wood pigeons that zoomed about
his estate. Each year he turned loose 300 crows (that had
survived the pot) in an attempt to repopulate the surrounding
countryside. The birds, Waterton gleefully wrote in a diary
entry, 'are no doubt considered a dangerous lot of rascals by the
good folk of this neighbourhood'.

Even in his twilight years when, in Waterton's words, his
hair was 'a shade as if exposed to the November hoar frost', he
still roamed his land each day, scaling oak trees barefoot and
spending hours monitoring the birds that had settled there. His
other priority was staking out poachers and rats that invaded
his land, the latter he loathed with a passion which bordered
on an obsession.

In his mission to wipe out every last rat from his estate, he
left out concoctions of porridge and treacle laced with arsenic,
and recruited a team of ratters spearheaded by a wild margay
cat collected from his travels. On one occasion, the otherwise
gentle Waterton was seen holding a rat by its tail and dashing
its brains out with a cry of 'death to all Hanoverians'. 'When I
am gone to dust,' he wrote to a friend in 1839 in letters which
are compiled in the archive at the British Library, 'if my ghost
should hover o'er this mansion, it will rejoice to hear the
remark that Charles Waterton effectively cleared the premises
at Walton Hall of every Hanoverian rat. Young and old.'

A year or so before his death, Waterton was described as
'looking like a spider after a long winter'. The modern day
naturalist Gerald Durrell preferred a different description:
'His life story reads like something invented by Edgar Allan
Poe with a certain amount of help from Richard Jefferies.'

Charles Waterton certainly shared one character trait with
Poe: a special affinity for the raven. Coming from a family of

Catholics, who had traditionally faced all manner of persecution on their West Yorkshire estate from the Reformation onwards, the Squire of Walton Hall possessed great sympathy for the underdog. The demise of the raven that had been wiped out from West Riding in his lifetime resonated deep within him as a great tragedy of his era. The last buzzard in the area was shot dead in 1813. A few years after that, Waterton's neighbour, Sir William Pilkington, confessed that his gamekeeper had shot the last raven in Yorkshire. The Squire of Walton Hall denounced him as a scoundrel (although according to his diaries they later ended the quarrel amicably).

Waterton devoted an entire chapter to the raven in his 1839 book, *Essays on Natural History*, where he described the final nest near to his estate being pillaged: 'the poor female shot dead to the ground' and the male 'deserting us for ever'. Waterton concluded his essay: 'Pity it is that the raven, a bird of such note and consequence in times gone by, should be exposed to unrelenting persecution in our own days of professed philanthropy.'

In the early 1830s, Waterton travelled to Flamborough Head, on Yorkshire's east coast, whose chalk cliffs are still renowned for their bird life. There, he struck a deal with a village trader to buy a young raven. He named the bird Marco.

'Marco could do everything,' Waterton wrote, gushing that he was 'as playful as a kitten', demonstrated 'vast aptitude in learning to talk' and could even mimic songs. Whenever a carriage approached Walton Hall, Marco would fly down to the bridge to greet it, and later fly back alongside the vehicle until it had left the estate.

Waterton would often observe his raven playing, sliding on its back down heaps of sand or snow. It is a game witnessed by many modern raven enthusiasts, although none have been able to give me a scientific explanation as to why, aside from the simplest and most obvious hypothesis – that the birds do it for fun.

On occasion, Waterton was confronted by the dark side of his raven, 'his evil genius prompted him to commit almost unpardonable excesses'. One day, he wrote, Marco 'took a sudden dislike to an old duck, which until then he had been on the best of terms'. With a stroke, the raven slashed its beak down and killed her instantly. When Marco had committed certain wrongs, Waterton was prone to repeat the same scolding rhyme to his bird: 'In all thy humours, whether grave or mellow / thou art such a testy pleasant fellow / hast so much wit, and mirth, and glee about thee / there is no living with thee or without thee.'

Inevitably, as with his wild Yorkshire cousins, Marco's demise finally came about at human hands. One day the raven decided to peck the estate coachman hard on the thumb, prompting him to strangle the bird on the spot.

Waterton never replaced Marco, but continued to grieve him for years afterwards. 'No bird in creation exhibits finer symmetry than the raven. His beautiful proportions and glossy plumage are calculated to strike the eye of every beholder with admiration. According to our notion of things, no bird can be better provided with the means of making his way through the world: for his armour is solid; his spirit unconquerable; and strength surprising.'

Waterton, a notably concise writer for his subject and era, would have chosen these words carefully. The raven's own lofty perch in the natural order means no other predators possess any real threat – aside from humans. In the West Riding of Waterton's time, they came at the raven with poison, spring-loaded traps, nets and bullets; weapons for which the raven's armour would never be solid enough. The raven's demise here was as conclusive as anywhere in the country. When Pilkington's gamekeeper squeezed the trigger and a puff of black feathers exploded in the sky, its total decimation was complete.

★ ★ ★

The progress of man is measured out in the species we have laid waste to. When the ice sheets melted across Britain around 10,000 years ago, Mesolithic hunter-gatherers set out across these virgin lands with spear, bow and arrow in hand. They killed auroch, wolf, lynx, brown bear, wild boar, beaver and elk, which were all once prolific across the great forests and wildwoods. The UK's bison, elk and brown bear were wiped out by 500 AD, and the last wild wolf in Britain was supposedly killed by a Highland chief called Sir Ewen Cameron of Lochiel, in 1685. Our appetite for destruction kept the ravens close. They knew that human footsteps would always lead to blood. Yet in the centuries that followed, our attention turned to ravens themselves.

The Preservation of Grain Act passed in 1532 by Henry VIII and strengthened by Elizabeth I in 1566, made it compulsory for every man, woman and child to kill as many creatures as possible that appeared on an official list of 'vermin', in order to protect crops and livestock. Bounties for the bodies of vermin were administered by churchwardens. One could expect to be paid four pence for bringing the head of a raven, kite or jay. Kingfishers were valued at one pence a head; so too a clutch of six young crows.

The extent of the animals and species that were slaughtered over subsequent centuries is enough to cause the modern reader a sharp intake of breath. On the hit list was most of our best-loved animals: hedgehogs, otters, frogs, and all manner of songbirds. Roger Lovegrove, the former director of the RSPB in Wales whose book, *Silent Fields*, is the most authoritative examination of the Vermin Acts and subsequent culls, estimates that between the eighteenth and the mid-twentieth century as many as 100 million house sparrows and their eggs could have been deliberately killed. I once met Roger at his home near the Welsh borders. It was a late spring day, with songbirds dancing up the creepers outside of his house and flitting between the feeders in his beautifully kept garden. Inevitably, our conversation shifted to the raven. For the first time in a long while he had heard them over his house again, he told me delightedly. As we sat sipping tea, I strained my ears to hear one croaking past the study window.

As Lovegrove points out, the backdrop to the Vermin Acts of the mid-sixteenth century was an increasingly crowded island struggling to feed itself. Everything in the landscape was subjugated towards our own development. The scientific basis for the killings was patchy at best: the hedgehog, for example, was mistakenly blamed for suckling the teats of livestock at night, thereby draining them of their milk. But while the reasoning may have been far-fetched, the act of killing itself was conducted with brutal efficiency.

In his book, Lovegrove pays particular attention to the demise of the raven. Throughout his research, poring over more than 1,000 parish records, he discovered what he calls 'a clear picture of the intensity and geographical distribution of the killing of ravens'. By the end of the seventeenth century, the raven had already been driven from the lowlands of England and Wales. The methods used by trappers included lime sticks (placed around the corpse of an animal to which the ravens became stuck) and smearing corpses with nux vomica, the toxic seed of an East Indian tree from which strychnine is derived. In his 1768 pamphlet, *The Universal Directory for Taking Alive and Destroying Rats and all Other Kinds of Four-footed and Winged Vermin*, the trapper Robert Smith boasts of the number of ravens he was able to kill in any single day. They were often the first to arrive at any new food bonanza where he was lying in wait.

Such sustained pressure quickly drove ravens from the English home counties. The uplands, where the dwindling few birds remained, proved scant sanctuary. According to Lovegrove's research, in just a few years during the eighteenth century, more than 5,000 ravens were killed in the Lake District. In Wirksworth in Derbyshire, 1,775 individual birds were exterminated between 1707 and 1725. Even small communities like Slaidburn in the Forest of Bowland recorded bounties for the bodies of 130 ravens killed between 1681 and 1728.

At the same time, the Parliamentary Enclosure Acts were beginning to carve up the countryside between landowners, and whole estates were turning over their land to the preservation of game, at the expense of anything else. This meant that even as the numbers of vermin killed started to flatten out at the turn of the nineteenth century and into Charles Waterton's era, the raven became more hemmed in. In East Anglia, which saw some of the most intensive game preservation in the country, the raven was presumed extinct by 1840.

By 1874 the raven had disappeared from the Welsh borders all the way to the east coast of England. At the same time, a quarter of Britain was held in estates of more than 10,000 acres. In this new cultivated and divided country, there was seemingly no place for the raven.

After meeting Roger Lovegrove and consulting on his methods, I dig out the old churchwarden's records of Charles Waterton's estate to see which animals were killed under the Vermin Acts. I study the records for Wakefield St John the Baptist and Wakefield All Saints from the mid-eighteenth century onwards, as well as St Peter and St Leonard's Church in the village of Horbury, not far from Walton Hall. I also look at St Helen's in Sandal Magna, where, in the middle of the fifteenth century, the Waterton family built their personal chapel. It was here that in 1830 Waterton's young wife Anne was buried in the family vault after she succumbed to sepsis following the birth of their only son, Edmund. I am also helped by the Wakefield Naturalists' Society, who provide me with supplementary evidence to my own research of the vermin destroyed.

The churchwarden's records all emerge from the archives in cracked, leather-bound volumes, with each single outgoing for any particular year recorded in painstaking detail. There is money for items like baskets, broom heads and sacks of coal; ale for the local bell-ringers and communion wine for the vicar. In between these seemingly gentle daily rhythms of a bygone way of life are the vermin payments. One barely notices them at first, and then they start to leap off the page. Foumarts (as polecats were known) are four pence for a body and one for a head. Hedgehogs, bull finch, sparrows and otters are two pennies apiece. Some entries are simply for unspecified 'vermin', although the 12 shillings paid out gives some hint at the number of animals slaughtered.

Despite several days poring over these records from the mid-1750s onwards in and around Wakefield, not a single entry for the raven shows up. I go back to Roger Lovegrove and ask him why he thinks this is, and he too says he failed to find many entries during his research in the West Riding. His explanation is simple; by the time they started counting, most of the ravens had already gone.

* * *

As part of my research in the Wakefield archives, I came across a small book, written in six short weeks, by pupils from St Peter's Church of England School in Horbury. The book was put together in 1968 and overseen by a seemingly visionary teacher called Alec Hinchliffe. In a time before strict national curriculums, Mr Hinchliffe had involved a range of folklorists and writers to help his pupils put together their book. Among them, was the then emerging children's fantasy author Alan Garner, whose novel *The Weirdstone of Brisingamen*, had recently been published and was fast becoming a favourite in Britain's schools.

Garner's work would go on to be recognised as one of the great pieces of fantasy writing of the twentieth century, and one of my own childhood favourites. The pupils' book, produced with his assistance, was intended in a similar vein and called, *A Time of Ravens*. I first began reading it as a distraction from the endless reams of parish records, but I soon became hooked. It is set on the North Yorkshire coast, amid the old seaside towns of Ravenscar and Staithes, both of which are isolated, spectacular villages full of intrigue and history and both of which I know very well. Staithes, in particular, I have a long connection with and have spent many happy nights in a seaside cottage owned by my dad's best friend, lulled to sleep as the great North Sea heaves outside my bedroom window.

It is a place where history is layered, like the fossils in the old sandstone cliffs, and the streets twist and turn in labyrinthine patterns between tumbledown cottages. Winter is my favourite season to visit, when the salt spray blasts so fierce it feels like you shed a layer of skin each time you step outside, and coal smoke hauls at right angles off the chimneys. In this evocative setting, the Horbury pupils conjured a story of gypsies, boggarts and ghosts of Viking warriors marching alongside the old, Victorian alum quarries. And throughout the narrative, slipping between the centuries is the form of the raven.

The children do not shy away from the bird's violence, imagining an urgent meeting at the local village hall, where farmers insist that the local raven population must be killed to stop them attacking newborn lambs. When I read that, it made me think of the desperate Caithness crofters I met during lambing season.

As the Horbury story progresses, the raven begins to transcend the physical. The Viking ghosts wield raven banners, and the gypsies deliver raven curses. Much like the Norse and Celt legends I have read, the bird is seen as a symbol of protection against death as well as an omen of foreboding. It is through a raven amulet that the heroine of the book manages to fend off a witch.

As I read, I become increasingly intrigued by the children's response to the raven, and how the dichotomy of the bird seems to encapsulate so much of its place in human history. In his prologue, the teacher, Alec Hinchliffe, describes the process by which it was written, with different class members working through plotlines in groups and driving the narrative themselves. 'There was initially no plot other than the fact that the children knew how the story was to begin and end,' he wrote. 'The descriptions of the raven and its importance through the book was all the children's own doing.' This seems to me all the more remarkable since most of these youngsters would have probably never seen one.

For Mr Hinchliffe, his project exposes the 'true nature of children', with the book they created the product of their own deep 'dreams, imaginations and nightmares'. Also, he says, it unravels our complex relationship with lore and legend and why we seek symbols like the raven to give expression to our own hopes and fears – often inexplicable in terms of our own everyday world. 'It occurs to us that human aspirations will always outstrip man's ability to rationalise them,' Mr Hinchliffe wrote. 'Therefore, we must have within our own culture the "fetishes" by which we try to come to terms with the world'.

I did not expect to find it so well proven in a primary school text, but it is a theory that goes right back to the beasts of battle motifs of Old English, and why the raven has stalked our literature and mythology ever since. We seek to explain ourselves through the bird. Even as we dedicated centuries to blasting the raven out of existence, its presence still always meant far more to us than mere 'vermin'. It always will.

* * *

Charles Waterton knew this. Not just about the raven but all the wildlife that was vanishing around him. He banned his gamekeepers from shooting or trapping any birds, and over time managed to attract back an array of species that had disappeared from his land. He constructed a bank for sand martins and allowed the ruins of an old gateway to become entirely consumed by ivy to provide nesting nooks for barn owls. Eventually, it became home to 49 pairs of different birds. In the broken trunks of decaying sycamores, he built stone ledges for jackdaws and carved out honey fungus from ash trees for tawny owls to better conceal themselves.

The following spring, after he had built his wall around the estate, six pairs of herons arrived. Over time it grew to 43 pairs. Watching from a hollow by the lake in one of his four strategically placed cylindrical stone towers (early prototypes of the modern bird hide), he counted peewits and wisps of snipe rootling for worms along the muddy shore. The detailed notes he kept also show nightjar, corncrake, nightingale, mistle thrush, starling, falcon, hawk, starling and fieldfare. Once, at Walton Hall, he counted 160 rook nests and 5,000 wood pigeons in a single day.

As the coal mines dug deeper around him, and a soap factory sprung up next to his land, poisoning the local water supply and prompting him into prolonged legal action (which he eventually won), Walton Hall became a walled paradise protected from the ravages of industry. In a letter to a friend in 1849, he wrote how carrion crows, magpies, hawks and kingfishers had all enjoyed a good year on his land. 'They may thank their stars that they have my wall to protect them,' he said. 'But for it, their race would be extinct in this depraved and demoralised part of Yorkshire.' Despite Waterton's best efforts at conservation, the raven never reappeared around Walton Hall. 'His noble aspect, his aerial evolutions and his wonderful modulations of voice, all contribute to render him an ornament of any gentleman's park,' he once wrote. Not in his lifetime, though.

Before visiting Walton Hall, I get in touch with the Wakefield Naturalists' Society to inquire whether or not ravens are seen around the area any more. There has been a great success with peregrine falcons in the city in recent years, with the birds now nesting on Wakefield Cathedral. As ravens and peregrines often prefer similar sites to rear their young, I hoped for good news. Alas, I am told not. While a pair of ravens are nesting in a nearby quarry off the A1 and have been spotted over various nature reserves, there are no modern records of them being seen over or near Walton Hall.

Even if they had been, nowadays there is nobody like Charles Waterton to keep a constant eye over the avian inhabitants of the estate. When he died, it was passed to his son Edmund, but he had inherited little of his father's conservationist zeal and was forced to sell the hall and its surrounding lands to pay off his debts.

Nowadays, Walton Hall is a hotel and spa catering for weddings, conferences and the occasional funeral. Most of the land that Waterton so carefully cultivated has been transformed into an 18-hole golf course. According to his biographer, Julia Blackburn, the sand martin bank that he built, complete with 56 nesting holes, was broken up and used as ballast for a squash court.

Having read so many of Waterton's diaries with their descriptions of life teeming on the estate, the rain-soaked Thursday morning I visit makes a thoroughly depressing trip. It is late autumn, and the groundsmen are hard at work blowing leaves off the immaculate fairways; a low mechanical rumble follows me wherever I walk. There are a few birds to see: dunnock, robin, blackbird and a skein of Canada geese honking towards the water, but the wildness that Waterton curated has largely gone.

He never wanted to keep people off his land. Indeed, in the 1830s he opened his park and museum of taxidermy to the public. By early 1840, Waterton estimated that 17,000 people a year came to enjoy his estate.

Behind the modern spa building, I follow a stream to discover the remains of his grotto, an old stone temple and outhouses he built for locals to picnic in when the weather was fine. He wanted to instil a love of nature in Wakefield's residents so they too would protect it as fiercely as he. I find it has eroded down to its foundations, mostly buried under bramble, rhododendrons and flat-leaved ferns that are just beginning to unfurl.

I walk down to the lake and chat to an elderly fisherman in head-to-toe khaki with a poppy pinned to his cap, who is keeping an eye out for brown and rainbow trout rising from its depths. There is a small island out in the middle of the lake where cormorants have sprayed the upper branches of wispy birch trees white. A few of the birds stand there now, drying their wings, drawn black against the grey water like hieroglyphics scrawled on a temple wall. The fisherman notices my admiring glances towards the cormorants and tells me, with a smile, that they try to shoot the birds when they can obtain a licence as they pose a 'threat' to the fish. He assures me he is a crack shot and that he is employed on a part-time basis to shoot rats on nearby pheasant shooting estates. At least on this matter, the ghost of Charles Waterton would approve.

I ask him where I can find Waterton's grave and the fisherman points me down another fairway and into a patch of woodland. Away from the golf course at the far end of the lake, the tree canopy of birch, oak, willow and elm grows close, and one can still glean an idea of what this land was like in Waterton's time.

I squelch along the muddy path, passing scolding wrens as it winds beyond the lake and down into a creek where the water slides green and still. I spot the stately, silver figure of a heron poised by the bank. As I walk, I notice giant mushrooms nearly a footlong growing sideways off the base of tree trunks, blancmange-white against the forest floor. I search the upper branches of the oldest trees for signs of the nest boxes that Waterton built, but see none.

After stepping over a stream, as the fisherman instructed, I cut right into the trees. A few paces more, and then I see the granite cross by a rusting, wrought iron gate that surrounds the grave. The oak trees that used to stand sentinel either side of it have long been cut down.

I sit by Waterton's graveside on a mouldering log, feeling the cool, damp moss sprouting up at the base of the headstone, and tracing my fingers over the metal plaque where his epitaph is carved. After so long reading about his life, I have looked forward to visiting his final resting place. But it's a dispiriting experience to think of the man left lying here for ever, while every year that passes his old empire crumbles further into history. My mood worsens when I try to find the remnants of the wall he built, only to discover most of it gone and the floor littered with spent shotgun shells and old crisp packets. On the other side is arable pastureland where woods once presumably stood.

I do manage to find some sections where the wall still stands taller than my six feet and I peer deep into its crannies; spotting woodlice, spiders and the remnants of an old bird's nest woven deep within the crumbled stone, framed by vivid green lichen and waterfalls of ivy. Some solace that here, at least, is life that the old naturalist still gives a home to. On the floor by one such ruined section, I notice a two-inch piece of moss-covered stone that would have been used to fill the wall. They call these tiny pieces the 'heartings' in drystone walling terms, the very centre of the construction that holds the whole thing together. I pick it up and put it in my pocket, to take some small piece of Waterton's legacy away.

When I return to Walton Hall, the rain is heavy enough to bring the muddy waters of the lake to the boil. The funeral party has already moved on, and I sit with a coffee in the empty hotel café. It is now nicknamed 'Charlie's Bar' – after the teetotaller Waterton, in the depressing manner in which commerce misconstrues and misappropriates heritage. I ask if the old fireplace where he slept is still visible, and am told the whole building has been refitted, apart from the lobby's original wood panelling, which smells of air freshener and chlorine. Nowadays they have conferences where the Squire of Walton Hall once wrote his diaries by candlelight, and prioritise golf balls over birds flying over the country's first nature reserve.

I keep that fragment from the wall on a shelf by my desk, near my collection of Staithes fossils that I have found over the years. As the months pass, the moss remains surprisingly verdant, a dry softness like baby hair. The stone is swirled with different spirals of grey. When I am writing, I sometimes rub it between my fingers to better concentrate. Keeping that piece of wall in my study means I do not readily forget my visit to Walton Hall.

A few months later, I resolve to visit again, if nothing else then to explore the woods in better weather and give myself another chance of spotting ravens somewhere nearby. It is a Saturday in spring, and the golfers are out in force, reclaiming the land after the most severe storm of the year. When I turn into the woods, it is noticeable just how many trees have been ripped up from their roots by the wind and dumped back on the ground. Some split and toppled birch trunks are even blocking the footpath towards Waterton's grave. The coots are still there, and this time chiffchaffs and chaffinch are beginning to build their nests. Carpets of soon-to-open bluebells have unfurled from the earth that was covered by fallen leaves on my last visit.

I hear a woodpecker drumming a rhapsody that echoes deep within the old woods. It takes me an age to see it, between the still-bare branches of a mighty beech, until eventually, it shifts around the trunk it is hammering and comes into view. After a few more strikes it takes off and flies over my head in a flash of speckled scarlet. As it passes, I hear the air thrum through the perfect folds of its beating wings.

I cut off the path, past the stream and into the wood as before, but this time, I cannot spot Waterton's grave. At first, I presume I've got my bearings wrong and even double back on myself a few times. Eventually, after tripping through the undergrowth and snagging my jeans on brambles, I see the headstone. It has been obscured by a large conifer uprooted in the storm.

The trunk is resting on the wrought iron fence around his grave, and its outer branches dangle down towards it, almost shrouding the whole thing altogether. I try to haul the trunk off, but it is too heavy to move. And then I notice something resting on top of the granite cross. Somebody has left an ivy branch, bent and tied into a perfect crown. The leaves are fresh and the crown recently made; a simple symbol of pagan reverence and spirituality. It is a heartening sight to see the 'weary bones' of Charles Waterton remembered in a way that he would have surely approved of. A crown for the resting green king, even if his kingdom has long been usurped.

Being back here among the old woods as the recently fallen trees decompose into the humus and saplings emerge in their place, breathing the rich scent of bluebells sprouting as

they have each spring for centuries, it makes me realise that there are things imprinted on a landscape. Time can sometimes move far slower than progress pretends. Perhaps it is too easy to simply lament what has gone before, forgetting how quickly symbols of the past can return.

After my last visit to Walton Hall, I receive an email from Francis Hickenbottom of the Wakefield Naturalists' Society who has been helping me out with my research. It starts: 'Dear Joe, you must take a look at this' and contains a photograph of a raven, its flight feathers razor-sharp; wingtips splayed into six long fingers; the unmistakable bulk of its beak and tail silhouetted against the grey sky. The picture has been submitted to the group by a local birdwatcher, who spotted it flying over Anglers Country Park near Wakefield, less than a throw of my stolen stone from Charles Waterton's wall. Finally, it seems, the Squire of Walton Hall has his raven back.

CHAPTER ELEVEN

Living with Ravens

The first time I meet Igraine Skelton, it ends in a rescue. We had planned to meet one Friday morning in the grounds of Knaresborough Castle, where Igraine often brings her eight ravens to show to visitors. She has been Her Majesty's Keeper of the Knaresborough Castle Ravens since 2000, but unlike the Tower of London birds, hers is a semi-official position. She keeps the ravens in aviaries in her back garden in Harrogate and transports them in pet carriers like the ones you might take a dog to the vet in. Despite their popularity, Igraine and her ravens only made it into the official castle leaflet in 2010, and she still has to pay for her own parking, much to the irritation of her husband, George. This lack of support means that when something goes wrong, there often isn't anybody else around to help.

It is a grey, cold day and only a few tourists and dog walkers have made the trip up the steep hill to the old ruins of the castle. For centuries it has occupied a perch to the west of the town centre, overlooking the Nidd Viaduct. When I arrive, Igraine is instantly recognisable from the mock chain mail, skullcap and tunic she wears as part of her outfit. Known as the raven lady of Knaresborough, she is something of a celebrity in this part of Yorkshire. She even starred in an arthouse documentary: *The Unkindness of Igraine Hustwitt*

Skelton. The unkindness refers to the historic colloquialism for a flock of ravens – you would struggle to meet a more gregarious stranger than Igraine.

She is pointing out to a gaggle of visitors the different names and traits of her ravens, which are out of their cages and waddling about the castle grounds. The Tower of London birds are kept with their wings trimmed, but Igraine prefers to allow her ravens full flight and secures them instead with extendable dog leads hooked to the jesses on their ankles. One, Izabella, a fine bird with mottled blue and emerald neck feathers, sits looking out over the steep valley. Another raven on a bench has a white chest, which I have never seen before. Igraine tells the visitors he is Odo, an African pied raven. A boy in a bright anorak, no older than five, peers out from his mother's legs; terrified but clearly entranced by the birds around him.

Somebody asks Igraine a question about the castle and her attention is momentarily distracted from the birds. I absently watch Odo rootling around the bench, wiping his beak sideways off the wood as if sharpening a knife. He hops over to the handle of the dog lead, which is wedged firmly between the slats of the bench and begins tapping at the plastic casing with his beak.

I take a sip of coffee and look out over the Nidd Viaduct. It is a view so beautiful that on summer evenings when I worked as a local reporter in Harrogate and Knaresborough, I often drove up here alone, long after Igraine and her ravens had packed up for the night, and watched the great sandstone arches in the bridge reflect like shimmering Ts in the river. It is so high here that at sunset, all the rooftops and treetops of Knaresborough spill out beneath you; a patchwork of green, terracotta and gold.

The sound of the dog's lead falling on the floor snaps me out of my thoughts, and I look up just in time to see Odo escape into the sky, a few feet of lead dangling from his jesses.

Because of the extra weight he is carrying, his flying is laboured, and he has to gain altitude in slow, looping arcs. It is one of those collective moments of shock when all we can do is watch, in horrified silence, as Odo flies over our heads and flinch as the handle of the dog lead skims the tops of the trees. With a terrible inevitability, it soon gets snagged in one of the taller trees, and Odo twists down to earth like a spitfire shot from the sky; wings crooked and feathers blazing, calling out in a voice so shrill it sounds like screaming metal.

I clamber over a fence and crash down through the steep undergrowth to where the raven is calling. I find him hanging upside down more than 3m (10ft) up in the tree, the lead wrapped around the branches like a spider web. The raven's head is jerking around as he tries to get his bearings. The nictitating membrane across his eyes snaps back and forth. Odo looks wild and helpless in equal measure and I am afraid to get too close for fear of his beak slashing my arm.

Over and over again, he emits the same sharp, scared cry for help. When he hears Igraine calling out, as she makes her way down the hill and over to the tree, I see the most amazing transformation occur. He stops shaking instantly and his glittering eyes cloud over. As Igraine arrives and whispers soothing words, Odo hangs almost perfectly still as we clamber up and release him from his confinement.

Fortunately, the steep embankment means we can reach the upper branches without climbing the tree. When we eventually manage to untangle the bird, who seems entirely unharmed, even relaxed considering his ordeal, we head back up the path to the watching crowd outside the castle. There is a ripple of applause when we haul ourselves back up. I notice that alongside the humans, the other ravens are also lined up watching us intently.

★ ★ ★

An hour or so later we sit in Igraine's living room, drinking tea and laughing with relief that the escape attempt was not any worse. Igraine has boiled a kettle that she keeps next to a toaster by the sofa. It seems strange, as her kitchen is only the other side of the hall behind us, but I can hear ravens flying about and the occasional clatter of cutlery in there. 'It is just easier this way,' Igraine admits.

The kitchen birds are producing curious sounds. One has learned to quack like a duck (this is from when Igraine used to sing *Old MacDonald's Farm* to them). Another seems to be speaking English, barking: 'what's the matter, Mum?' I would normally be surprised at such linguistic feats, but earlier in the day on a bowling green next to the castle, one of Igraine's ravens, Mourdour, had greeted me in a perfect Yorkshire accent, asking: 'are you all right?'

In her living room, there are two sofas and a television. A drawing from the York Mystery Plays hangs on the wall. It was commissioned by the director as a thank you gift to Igraine and another of her ravens, Gabriel, who was asked to appear one year despite fears that he might eat one of the doves. There is also a painting of the aircraft carrier HMS *Ark Royal* on which George, a retired plumbing and heating engineer, once served. Two juvenile rooks perch on the edge of a cot by the window. Little attention is paid to them by Igraine's cats, Sylvester and Molly, who are well used to corvids muscling in on their domain.

It is a few weeks before Christmas, and on the mantelpiece are a number of advent calendars being slowly pecked open by the ravens. They are only allowed half a chocolate each day, Igraine explains, because of a substance in cocoa that can be poisonous to them. For the same reason, she lets each of her ravens crack open an Easter egg, although the birds are banned from eating what's inside.

I wonder how someone ends up living in such close proximity to ravens and why anybody would want, not just to share their home with the birds, but relinquish so much of it to them? Over the course of writing this book, I have asked these same two questions of an entire cross-section of people around the country who have taken the decision to live with ravens. Some keep them in aviaries in their gardens; some give over whole rooms to the birds. Igraine does a little bit of both.

The reasons given by Igraine and others are as varied and complex as any attempt to explain why humans would want to live with any animals, or even members of our own species. The raven's humour, intelligence, mischievousness, protectiveness, companionship and even its ability to mourn, all result in people forming deep, and sometimes inexplicable, bonds with them, that go far beyond the simple notion of a pet. 'I tell people I live with the ravens,' Igraine says, 'not the other way around.'

★ ★ ★

Igraine was born 60 years ago and her first name is actually Janette – after Janette Scott; the English actress and star of *The Day of the Triffids*, whom her father was sweet on at the time. Her mother chose Igraine as her middle name, after the mother of King Arthur. As a child she was brought up with the old Arthurian romances, in which the raven appears so prominently.

When she was a girl, Igraine's grandfather worked as the head groom for Lady Collins at Knaresborough House, and he opened her eyes to the close bonds people can have with animals. When he died, her grandmother moved his collection of old taxidermy cabinets into a council house. Among the stuffed birds were a peregrine falcon and a tawny owl. Igraine fantasised about having a bird of her own one day; a real life one whose amber eyes would not stare out blankly and unblinking from behind a pane of glass.

Later in life she worked as a nurse at Harrogate District Hospital where she met her husband. George had two young sons from a previous relationship, Gary and Mark, and always one to keep the old fairy stories in mind, Igraine says she worked hard to not be seen as 'the wicked stepmother'. When Gary came home one day with a school assignment to write a story encouraging people to visit Knaresborough, she was more than happy to help.

Together they collated all the prominent landmarks of the old medieval and spa town: Mother Shipton's Cave, a petrifying well once said to be home to England's most famous prophetess; Knaresborough Castle, which dates back to 1130 and had been associated with the murderers of Thomas Becket and numerous monarchs; and the sandstone Chapel of our Lady of the Crag. They also decided to include in their story a raven who lived in a cave on the riverbank and guarded a precious gold coin.

Gary was a promising footballer, in spite of asthma, and dreamed of playing for Liverpool, his favourite team. But one Saturday he woke up on the day of a school cup game with a nasty cold. The doctor was called, and though Gary just wanted to find out if he could still play the game that afternoon, his condition quickly deteriorated. 'His trachea ended up swelling so much when he tried to breathe really hard it burst,' Igraine recalls. 'He drowned in his own blood.' It was 1986 and Gary was 11.

Later, to help with her grief, she decided to elaborate on the story they had started together, turning it into a novel-length book in his memory. And she decided to get a raven. It was 1999 when the first of her birds arrived from a breeder in Cumbria who advertised her in *Cage and Aviary Birds* magazine. The raven's previous owner had abandoned her after she had spooked his collection of mynah birds and stopped them breeding. According to Igraine, when she collected the unwanted raven she was a furious mass of feathers and beak, and on the drive home nearly tore her cardboard box into shreds. Igraine contemplated this raging bird, and decided to name her Ravenelf.

'At first we let her out in the kitchen. She stood on the larder door and for two days if you went near her she barked and was really aggressive. We thought, what are we going to do? We've got a mad raven here. But once they know you're scared that's it, you've lost them. So you can't let them know.'

As the days passed, hunger eventually got the better of Ravenelf and she started accepting morsels of food. They managed to put anklets and jesses on her legs and set up a temporary roost for her in a spare room in the house. The real moment of bonding came the following summer when George's Staffordshire terrier attacked Ravenelf, leaving her with a broken leg. George had recently been signed off work injured, after falling off a ladder and cracking his ribs. The 2000 Sydney Olympics had just started, and so bird and man sat recuperating on the sofa in front of the television, cheering the games as their respective broken bones slowly meshed back together.

'She imprinted on to him after that,' Igraine says, 'and it changed her whole being for the better. She realised we weren't going to hurt her. It was fascinating to watch. Like a personality transplant. She became completely different.'

Her story of the transformation of Ravenelf reminds me of Loki, the raven I met on the farm in the Lea Valley who had similarly bonded with his owners after at first lashing out against them. Of course, almost any animal needs a period of adjustment after arriving in a new domain, but there is something about the pattern of relationships ravens form with humans that fascinates me. A hot blast of fury that cools into an eternal bond, the ensuing love just as fierce as the initial anger. It gives us some insight into the length of time it takes pairs to bond and breed in the wild. Juvenile ravens will take several years after forming a pair before they are ready to rear their young. When you watch them together, marking their domain or soaring over a crag, it is remarkable to see the tenderness often on display. Left to their own devices, ravens will stay together for life.

Ravenelf is 18 now, and Igraine admits she does not have long left. After her broken leg healed, she developed a condition known as bumblefoot, common among captive birds, where the bone calcifies into solid lumps. It means she can't easily bear her own weight and has largely stopped coming to the castle as part of Igraine's troupe. She still speaks from her kitchen cage, asking: 'what's the matter, Mum?' and declaring: 'I'm not a raven, I'm a crow.' Her words rile another raven perched in the kitchen, sending it hopping from cabinet to cutlery drawer with a leathery flap of wings that appear disconcertingly huge within the confines of the room.

I rest my finger on the bars and Ravenelf moves her beak towards it, but I lose my nerve at the last moment and pull away. She is picky about who she will let touch her, Igraine tells me. 'She is very good at knowing who is pure of heart.'

We step outside and into an aviary under an apple tree. Inside, two ravens, Mongo and Viviane, growl deeply at our arrival and leap from perch to perch animated by the stranger in their midst. I feel two sets of gleaming eyes upon me as the birds dart over my head. They have been paired up for several years and in the recent breeding season surprised Igraine by building a nest and producing a clutch of eggs. She feeds the birds pheasant and unskinned rabbits that she buys in bulk from the butchers. They are magnificently glossy as a result.

While wary of me, Igraine's ravens are incredibly tactile around her, submitting their heads to be stroked. Sometimes, she says, they will go so far as to peck her hand when she stops preening them. Like a crudely sketched shadow, the birds all flutter after her as she goes about her house, always aware of what rooms she is in, fixed to her by invisible bonds.

Even with their master, though, they can prove capricious, and she says the worst thing about living with ravens is their unpredictable mood swings. At the castle, they are often downright mischievous. The ravens will sometimes play injured to distract the attention of visitors, pretending for example to have their foot snagged in a bottle and when somebody comes to help, they suddenly spring up and try to pilfer whatever's in their bag. In their hiding place on top of the castle keep, Igraine has found mobile phones, digital cameras, babies' dummies, clothes and shoes. According to the local papers, Izabella, who Igraine has had for a decade, was being considered by the local constabulary for an ASBO after she learnt to say, 'what the fuck are you looking at?' to visitors.

'Centuries ago, Ravenelf, Izabella and Mourdour would all be shot because they use human speech,' Igraine says. 'One of the reasons why I wanted to do this was to show the playful side of ravens. They can have people in stitches.'

She is fond of dispelling myths about the birds, and showcasing their intellectual and emotional abilities. While happy to answer any questions, the one word which Igraine always takes serious umbrage to is pets. 'That makes me really angry,' she says. 'They are not my pets; they are my companions.'

★ ★ ★

The difference between a companion and a pet, I suppose, is what one receives in return. Generally, with a pet, we hope for unconditional love, comfort or, at the very least, an appreciation of beauty or satisfaction of biological interest. Dog or cat owners may well disagree with me, but a

companion is something different. Companion implies a
relationship where you share emotion, knowledge and
intimacy. Those like Igraine, who live with the birds, say
there is something soulful within the raven that connects
with humans today, just as deeply as it has done for centuries.
It inspires the feeling that we can learn from the bird, and
that it understands us. We keep pets and yet we live with
ravens. The bird is a prophet, a mischief-maker, a killer and a
confidant.

One of the best literary representations of this relationship
was written by Charles Dickens. On 28 January 1841,
Dickens wrote to a friend, George Cattermole, detailing
plans for his new novel, *Barnaby Rudge*, the first chapters of
which were soon released to the public as a weekly serial.
'My notion is to have [Barnaby] always in company with a
pet raven,' Dickens said, 'who is immeasurably more knowing
than himself.'

Dickens could describe the bond we share with ravens so
keenly because he experienced it for himself. He kept two
ravens, both called Grip, who were inspiration for the bird in
Barnaby Rudge. The first slept in a stable, often on the back of
a horse and took delight in tormenting a pet Newfoundland
dog, stealing the unwitting creature's dinner from under his
nose. According to the preface to *Barnaby Rudge*, one day the
stable was newly painted and the workmen went off for their
lunch leaving behind a tin of paint. Grip ate a full pound or
two of the stuff, and swiftly dropped dead from lead
poisoning. Dickens described himself as 'inconsolable'.

Grip (version two) was an older raven, which a friend of
Dickens had spotted at a local pub where it had acquired the
skill of chasing out drunks at closing time. When Dickens
owned the bird, it proved such a talented mimic that it could
sit on the author's window and drive horses out of the stable.
Despite its destructive tendencies, digging out mortar from

walls and smashing panes of glass by removing all the putty,
Dickens admitted he could not have had more respect for the
bird. Sadly, it only attached itself to the cook and scorned
every other member of his household, including Dickens
himself. After a few years, Grip 2 also died suddenly after
consuming an unknown poison. It collapsed while watching
a piece of meat roasting on the fire and uttered the final
plaintive cry of 'cuckoo' as its last word. As befitting the time,
Dickens had his raven stuffed by a taxidermist and kept the
body in a case by his desk.

The second incarnation of Grip seems to have informed
the character of the same name in *Barnaby Rudge*. The
fictional Grip is a multi-talented bird – shrieking 'I'm a devil',
waddling like a gentleman in breeches and performing tricks
in exchange for tossed lumps of meat – but Dickens also has
the raven act as a spiritual guide to Barnaby through the
novel. As the protagonist reflects of his raven at one point:
'he's the master, and I'm the man.'

The raven was a popular pet in Victorian households.
According to Dickens, he was inspired to write about his birds
because of the warning from fellow Victorian man of letters,
Charles Waterton, that the raven had been wiped out in the
West Riding and would soon become extinct across England.

Four years after *Barnaby Rudge* was published came another
artistic depiction of the raven, which more than any other
modern piece of work has cemented the bird in the British
psyche as a harbinger of doom. Edgar Allan Poe was inspired
to write his famous poem by Dickens. The author had met
Poe during a trip to the US in 1842 and both the man and his
love for ravens clearly left an impression. In his review of
Barnaby Rudge, Poe describes the character of Grip as
'intensely amusing'.

Both writers are preoccupied with the raven as a
manifestation of the human soul, but if Dickens encompasses

its many conflicts then Poe plumbs only its bleakest depths. His is the very embodiment of helplessness and hopelessness; the impotent agony of mourning. Poe's raven haunts his pages, with only one word emanating from its beak: 'Nevermore'.

The success of Poe's poem meant that the many, often contradictory, viewpoints of the raven through the ages became crystallised in modern western culture as simply a bird of horror. If you have ever been to the Courtauld Gallery in London, you might have seen Gauguin's 1897 painting *Nevermore*. The artist painted a raven in the top left of the portrait looking down over the exposed body of a nude woman, describing it as 'the bird of the devil'.

Directors in twentieth-century Hollywood excitedly seized upon the bird as a lazy narrative representation of evil. My favourite, most overly hammed-up version, is the 1935 Bela Lugosi film, *The Raven*. Lugosi plays a mad surgeon with a torture chamber in his basement and – like Dickens – a stuffed raven in a bell jar on his desk, whose foreboding form the camera regularly zooms in on. The great director of the 1930s and 1940s, Frank Capra, also made allegorical use of ravens in his films, not least in *It's a Wonderful Life*, where Uncle Billy keeps one as a pet.

Uncle Billy's bird was played by a raven called Jimmy, the star of hundreds of other Hollywood films of the era – such was the demand among directors for corvids. James Stewart recalling the filming of *It's a Wonderful Life*, called Jimmy the Raven 'the smartest actor on set'.

Authors of horror and ghost stories still lean on the raven as a narrative crutch. Though in recent decades, the raven does seem to have undergone a bit of a cultural rehabilitation. J. R. R. Tolkien was the first to paint a more sympathetic portrait of ravens. In *The Hobbit*, Ravenhill, one of the foothills of the Lonely Mountain, is guarded by wise and friendly ravens. Chief among them is Roäc – 153 years old, blind, nearly bald and unable to fly. He directs his birds to carry secret messages bearing news of the mountain and advise travellers passing through of the dangers ahead.

In the current television series *Game of Thrones*, where most of the human characters are preoccupied with lust, power and violence, ravens are used as messengers between clans and castles. I find it interesting to compare these modern-day depictions of ravens to the ancient stories I have read and heard. Once more, it seems our understanding of the raven is becoming far more nuanced and ambiguous, more in line with the impression Dickens, rather than Poe, intended. There is a growing acceptance that the birds can be both savage and soulful.

* * *

In the summer of 2016, Steve Burns stood in the dock of Reading Magistrates' Court and told a raven story of his own. The 58-year-old had been arrested the previous year following a police raid on a bird sanctuary he runs from his home in Crowthorne, Berkshire, called Raven Haven. Officers discovered a cannabis factory housing nearly 70 plants and with 9kg ready to sell. They also seized 180 birds, many of which were put down because they were in such poor health. Throughout the police investigation, which was widely reported in the local and national press, Steve – who received a community award in 2012 for 22 years of providing sanctuary for injured birds at Raven Haven – refrained from speaking in public. He wanted to wait until the trial to explain his relationship with ravens.

The first wild raven he ever saw was in 1989, on the sands of Burton Bradstock beach in Dorset. It was the first day of a fortnight-long family camping holiday and the weather was unseasonably foul. As Steve, his wife and two children walked along the beach his daughter, Nina, noticed a black shape lying among some rocks. She approached with her brother Ben to investigate, and a huge beak reared up to warn them away. It was no more than a bundle of feathers, really, with a right eye missing, right leg partially severed and one wing almost entirely snapped off, but Steve recognised instantly what it was. He picked the raven up, wrapped it in a towel and carried it back to their tent.

Painstakingly, using the same methods by which he used to rescue crows in his youth, Steve washed the grit from the emaciated frame of the raven, bandaged his wounds and sat up most of the night trying to get him to drink some water. In the morning, he telephoned the RSPCA and RSPB but the only advice he claims to have received was to put the bird out of its misery. Instead, he took the raven to a local police station in Bridport who referred him to a specialist vet who stitched up his wounds and injected him with a barrage of antibiotics, all for the princely sum of £1 plus VAT. Following the operation, Steve was told by the vet that the raven was already between 65 and 70 years old (although this seems a rather drastic over-exaggeration) and could never be released in the wild again because of his injuries.

They named him Tarquin, and for the rest of the holiday the raven stayed with them in the family tent, eating mincemeat and dog food and slowly flickering back into life. When the fortnight was over, they decided to take Tarquin home. Steve built an aviary and kept him there alongside two mynah birds, called Austin and Morris, glossy starlings, quails and cockatiels. Although the raven was at the top of the food chain, he never attacked any of the other birds and would even wrap his wings around the quails at night to keep them warm. In return, the mynah birds preened Tarquin's feathers for mites. The raven also refrained from eating a goldfish in the aviary pond. Whenever Steve was around, Tarquin would perch on his shoulder, eye to eye with his saviour.

After a few years of living in harmony, Steve's then wife Shelley decided to find Tarquin a mate. They bought a captive female and introduced her into the aviary. Within a few moments, she killed the cockatiels as well as one of the mynah birds and the goldfish. 'She was like lightning,' Steve says. 'She wanted to mate with him and there was no way she was going to tolerate him having any friends whatsoever. I had to remove all the other birds sharpish.' After five years they had their first babies. Before his death in 2012, Tarquin fathered 23 ravens. When he was gone, Steve says, he lost his 'best mate'.

Tarquin's life story is inextricably interwoven with Steve's. His relationship with Tarquin was the reason he chose to set up his sanctuary, retire from his job as a telecoms engineer and devote himself to the ravens. It also demonstrates his refusal to give up hope for a bird regardless of the severity of its injuries. This was a factor that officers involved in the investigation claim led to the situation at the sanctuary getting 'out of control', in spite of its good intentions.

At the end of his three-day trial, Steve was found guilty of 25 wildlife offences and received a six-month suspended prison sentence as well as being ordered to do 60 hours unpaid work and pay £1,080 in costs. For the cannabis plants, which he says he was selling to raise funds to fit out the cages in the 'intake and recovery room', he received a further 18-month suspended sentence. To the dismay of two former volunteers at the sanctuary, who had reported him to the police, Steve was given a prohibition order preventing Raven Haven from taking in any new birds for a year. They had hoped he would be banned from keeping animals for life.

I had heard of Raven Haven but never spoken to Steve before the police raid, however the case interested me for obvious reasons. I phoned him one night before the trial and introduced myself. He sounded weary and wary, particularly because I said from the outset that I was a journalist in my day job. At a previous paper, I had worked as a crime correspondent and have sat through countless trials and sentencing hearings over the course of my career. But I assured Steve I simply wanted to hear his side of the story, and understand why a man who loved ravens as obviously as he did, could end up being prosecuted for animal cruelty. I had never been to visit the guilty party following a trial, always the victim. I also wanted to meet his ravens.

I discovered later, that the police hadn't taken away any of his ravens, only pigeons and other species of bird. His ravens had been impounded in their aviaries outside, with only Steve and the three existing volunteers who work at the

sanctuary allowed to feed them for the duration of the investigation. During the raid, he told me, the police were too intimidated to go into the raven aviaries for fear they would be attacked. Eventually, after a year or so of waiting for the case to come to a close and settle down, Steve agreed to let me visit.

★ ★ ★

Raven Haven is halfway between the elite boarding school Wellington College and Broadmoor high-security psychiatric hospital. It is set back from the main road and immediately identifiable from the plush neighbouring homes by a moulding caravan in the drive. Steve greets me at the front door wearing skinny jeans, silver high-top trainers, an eagle T-shirt, leather waistcoat and a pendant in the shape of a soaring raven hanging from his neck. He leads me past the living room, where a pigeon roosts on the dresser and the body of a stuffed raven stares at me as I pass, and into the kitchen.

In one corner of the room is a caged African grey parrot called Angelo, and in the other, by the patio door, is a magnificent blue-and-gold macaw named Sally. Steve sits at the kitchen table drinking from a mug that says 'sexy pants' and rolling a giant cigarette with a bamboo mat, in order to pack the tobacco in more tightly. When he speaks, it is in a broad Estuary accent and loud enough to counter the clatter of a washing machine and the croaking ravens in the aviaries outside. It is nesting time and several of his pairs have already laid clutches of eggs. On his forehead is an angry-looking scratch where one of his ravens pecked him, earlier that day, to keep him away from the nest.

After a few minutes chatting, we are joined by one of the volunteers who lives in the sprawling house, which Steve says was virtually derelict when he first moved in. Alex Falck-Doessing is a 23-year-old with a butterfly elaborately tattooed on her chest. Perhaps it is the product of her Danish Viking heritage, she says, but she adores ravens and has worked at the centre since bringing an injured pigeon here in her late teens. She was here the day of the police raid and says the experience was 'the worst day of my life'. Steve seems keen to put it all behind him, and describes it only as 'pretty grim and very distasteful'.

He pauses and takes a deep drag of his cigarette. 'In 25 years, we never turned away a single bird,' he tells me. 'We're a rescue centre. We've fixed thousands of broken legs and wings, prosthetic beaks and all kind of things. They were putting birds to sleep on the spot. Not a single one of them needed to be.'

He tells me the story of one particular rescue. A pigeon called Yoohoo was brought to him with its entire crop split open and cheek slashed apart following a botched sparrowhawk attack. Its insides were falling out of the open wound and he could hear the acids of the pigeon's vital organs hissing and fizzing. Steve tells me how he superglued the bird's crop back together and even managed to seal the flapping wound on its cheek. The bird, he says, made a full recovery and still regularly comes back to visit him.

Is it cruel, I wonder, to be so hell-bent on bringing birds back from the brink regardless of their injuries? Had he lost control of his duty of care to the birds, as the prosecution alleged in court, or was he simply refusing to write off animals that could still be treated? On this Steve is resolute. 'I will never let a bird suffer. If they needed to be, I would have taken them to the vet.'

Right on cue, a severely-disabled raven waddles into the kitchen. Her feet are turned in on themselves giving her the unsteady gait of a rickets sufferer. Still, she manages to hop up to a chair and on to the kitchen table, where she plonks herself next to Steve who tickles her throat feathers. At the sight of the raven, three others in an aviary outside start calling loudly and hammering on the patio door with their beaks.

Precious, Steve tells me, was born at the centre and suffered 'a near skeletal collapse' at only a couple of weeks old. 'We got her to the best we could,' he says, as Precious croaks with pleasure at being stroked. 'It took three months teaching her to walk, with me on my hands and knees in the garden. Then she couldn't stand on her perch without falling off. We had to stand for months just holding her. Then we taught her to jump. We've been to hell and back together.'

Steve tells me about another raven, Narnia, one of those flapping about outside. A few years ago, they were telephoned by somebody near Stonehenge and asked to take in a wounded bird. 'She had been shot with a shotgun and completely smashed up,' Steve says. 'She had six slugs in her, but survived on the ground for a couple of weeks. She had terrible infections: aspergillosis, bone rot, and full-blown septicaemia. She took herself to someone's doorstep and just sat there waiting for the man to get home from work. He opened the door and says she just walked in.

'But after she came here we set about fixing her up and she's a healthy bird now. We've had many birds through here that have been shot. It's gamekeepers and farmers without a doubt. That's something that always bugs me. They shoot everything that moves because they think they are there to eat the crops. That's why these birds are moving into towns where they are safer.'

Sure enough, he says, wild ravens now nest around Crowthorne. Part of the population may well be the offspring from birds that have escaped from Steve's aviaries over the years. Alex finds it remarkable that they still return to visit every year. In the past, she has come down on a morning to discover wild raven footprints, from what she believes to be the visiting escapees, in a fresh dusting of snow.

We walk out to the aviaries, past an incubator where a clutch of small green and brown-speckled raven eggs are waiting to hatch. They are smaller than I thought a raven egg would be, roughly the size of a Cadbury Creme Egg. The birds will hatch soon, cawing blindly and livid pink. As they grow bigger and their feathers begin to sprout, the inside of their mouths will stay rose-coloured. Slowly, as the bird reaches adulthood and sexual maturity at around three, the lining of its mouth will turn a bluish-black. Interestingly, dominant ravens can develop a black mouth far sooner than other birds, even a matter of months after hatching.

As I look at the eggs I think of the skeletons of adult ravens I have seen. There is one in the basement of the Horniman Museum in London, whose skull is the size of a small plum and beak extends out perhaps six inches from the lower mandibles, while its ribcage is as taut and delicate as a harp. Like those flapping about in the aviary outside, even in skeletal form the raven is a behemoth among birds.

Steve has 35 ravens along with numerous crows and rooks. The ravens, who all appear as large and seemingly healthy as any I have seen in the wild, are kept in compartmentalised aviaries that encircle the entire garden. In the middle is a pigeon house filled with clattering and cooing birds. Steve admits some of his newer neighbours on the street are less than impressed with his menagerie.

At the sight of the three of us, the ravens instantly strike up a cacophony of noise. 'They love human company,' Steve says. 'They just want to be with you. They will steal things just to get your attention.' A particularly favourite game is to steal his tobacco, pass it between them and empty it out in the water provided for them. They also enjoy getting his cigarette papers and ripping them out one by one. His wallet, too, often ends up getting a soaking, but never his mobile phone. 'They dangle it over the water and see my reaction and start laughing,' he says, 'but they've never actually dropped it in.'

Once, one of the ravens, Mystic, managed to steal the bunch of keys that unlocks all of the aviaries. She then managed to prise off every single key from the chain and hide them all over the garden. Steve had to saw all the locks off and replace them. Once that was done, Mystic started bringing one key back to him every day. Eventually, he ended up with the entire set returned.

Ravens, he says, still possess the capacity to surprise him with their playfulness and soulfulness, far more so than the parrot he keeps, despite scientific studies claiming they possess similar brainpower to the corvids.

The ravens will fight, sometimes savagely, among themselves. On 26 January 2007, the day of his daughter Siobhan's 14th birthday, he discovered the battered body of a raven called Spirit, which he believes was killed in a mob attack. 'It was a real tragic thing. It was coming up to the breeding season and we don't know what Spirit did to upset the others but it was obviously unforgivable and they killed her.'

Another time, one of the new volunteers at Raven Haven ignored warnings not to devote too much attention to one particular bird, as the others can become jealous. 'They all suddenly attacked this one raven at once,' says Steve. 'I had to go and rescue him and got quite a few holes in me. One went through my jeans and burrowed a good inch into the flesh in my groin. If that had been a few centimetres closer ...' and he trails off.

After hearing this, I am reluctant to accept Alex's offer to let me into one of the main aviaries, a wire mesh about 3m (10ft) by 1.5m (5ft), which houses four non-breeding ravens.

I have only ever been in a raven's aviary once before, and I am made wary from the memory of how quickly the birds can think and move. Still, I try to push any fear to one side, and step gingerly inside.

The ravens who until that point have been causing a racket of croaks and grunts, suddenly quieten at my presence. Three of the birds eyeball me from the far end of the aviary, while another flies back and forth over my head. I can feel the rush of wind from its beating wings over my scalp and hear its talons skitter over the wire mesh. The smell is that deep raven mustiness; the scent of decay and secrets the bird has buried. The remains of a yellow chick lie discarded on the floor. I am too intimidated by the pairs of black eyes upon me to step out from the corner of the aviary. The croaks grow in number and the nearest raven clatters closer and closer over my head. I don't even last two minutes before I scuttle back out the door: relieved and slightly ashamed at how easily the ravens assumed dominance over me.

Back in the kitchen over another cup of tea, I ask Steve what is the most curious thing he has discovered through living with them. 'They mourn,' he says. 'Which is really quite something. They show and suffer great grief.'

When Tarquin died, his partner Tilly began to self-harm, snapping all her primary feathers off, leaving her unable to fly. She only came out of her torpor after they introduced one of her sons into the aviary. Steve then tells me about another bird, Rox, a captive raven whose partner died four years ago after they had been together for 20 years. Following his death, she began to self-harm, severing her feathers deep down by the quill. She also stopped eating, Steve believes, because she wanted to commit suicide. Now Rox is improving, he says, but still each year when the breeding season comes around she tries to self-harm again. 'We're working hard on her,' he says, in the manner a psychiatrist might discuss a patient in nearby Broadmoor. 'It will pass.'

We compare the English view of the raven as a bird of death, to that of the Inuit and Native Americans where it is regarded as a mystic, a mischief-maker and a spirit guide. Steve fondles the raven charm around his neck and tells me he definitely subscribes to the latter. Like me, he says he has raven dreams, in particular one where he is flying in formation with his birds, arrowed like a skein of geese. But there is one story above all others that has convinced him of our connection to the birds.

A few years ago, one of his ravens escaped the aviary and for eight weeks there was no sign of it. Then he received a call from the manager of a nearby polo club to say that the raven had arrived on her land (she knew it belonged to Steve by the ring on its foot). The bird turned up not long after the club's head groom had drowned in a kayaking accident. His assistant had been with him, but survived. The assistant told Steve that the raven had turned up at his back door one day and sauntered into the house. When Steve came to collect it, the raven refused to leave, flying away from his car to a nearby tree and croaking an alarm call. It was only once the funeral had taken place, that it left the groundsman's side and flew off. 'He believed that the raven had come to take his friend's soul to the other side,' Steve says.

In the countless conversations I have had with people about ravens this is often what it comes back to; a feeling that the bird exists on a higher spiritual plain. In death and despair there is the sense that somehow the raven knows more than we do of the human experience, that for good or ill, this bird of darkness has arrived to tell us something.

★ ★ ★

A few months after I met Steve, I was walking through the grounds of Powderham Castle on the banks of the River Exe, interviewing the Earl of Devon for a newspaper piece. We got talking about ravens and the Earl told me that a pair had nested in the turrets of Belvedere Tower, a pseudo-medieval building in the grounds of the estate, which had been gutted by two major fires in the post-war era and was uninhabited, aside from the birds.

The Earl grew up in the castle and said the pair of ravens had been there for decades. He inherited the estate in 2015 following the death of his father, the 18th earl, at the age of 73. The day of the funeral, he remembers walking across the grounds of the estate and spotting one of the ravens fly over from the tower and begin circling an old cedar tree next to him. He told me he believes the raven was a sign, and that after seeing the bird and hearing its call, he was filled with the unshakeable feeling that the raven knew what had happened. He still cherishes that nest on the tower, and looks out for the birds every year.

Can it be true that when we are at our most raw, the birds seek us out to provide meaning to our own inexplicable lives? Or is it perhaps because this is the moment when we look beyond our selves to what is really there?

Igraine writes to tell me that Odo, the raven, has escaped again and this time he did not survive. He had tried to fly across a river, but ended up tumbling into the water. By the time Igraine reached him, it was already too late. When she lifted his limp head she noticed a white crust had formed under his tongue, the sign of a disease that, she believes, could have weakened him in flight.

I express my condolences and ask how the other ravens reacted to the death of one of their own. Igraine tells me that when she returned with Odo in her arms, two of the ravens, Izabella and Emyrus, both 'crowed loudly', an unusual call, but one she has heard before when a raven died. Mortimer, who shared an aviary with Odo, didn't realise he had gone and on the first evening of his death called out for him long after nightfall; the loud mournful croaks of a raven echoing down her suburban Harrogate street. The keeper and her ravens united in grief.

CHAPTER TWELVE

The Ravens in the Tower

Once, in an attempt to determine the exact age of my house, I gathered together nearly a century's worth of old Ordnance Survey maps of the area, dating from the late 1880s to 1961. We live in what is nowadays a busy suburb not far from Sheffield city centre, but back in 1888 – the earliest map I could find – neither my street nor any of the ones around it existed. The natural landmarks, happily, remained the same; the woods on a steep bank to the north where tawny owls still roost today; to the south, the River Sheaf, which has recently begun to clean up from years of industrial spoil and now has brown trout darting through its shallows. Yet back then, in between these natural borders, was nothing but farmland snaked with a few quiet lanes.

I flicked through the maps and across the decades, watching roads and houses spreading like roots through the earth, branching off from the main railway track to London, which runs along the middle of the valley. My own street, named like many in the area after one of the Yorkshire Dales, was built in the early Edwardian era, one house popping up at a time as if on a Monopoly board. By the 1960s, all that once-vacant space had been drained, sectioned off and piled up with neat little three-storey high brick rows of personal fiefdoms. Looking at this march of progress at such high speed, it seemed nothing short of an invasion of the land, one repeated in so many other mapped square inches of the country.

Over almost exactly this same period, as our urban areas tumefied – the ravens were obliterated. Much of my book has been about chasing ghosts, discovering the legends and the places where the ravens once were. Over the course of my travels, I have also built up in my mind a map of their return across every county. Sometimes it is anecdotal; the people who tell me they spotted a raven on a motorway barrier or flying over their house for the first time. Often, I have seen it with my own eyes; the ravens on the streets of Bristol, in the New and Kielder Forests, reoccupying the cliffs and crags and villages and towns that already bear their name, nesting on stately turrets and church spires; gliding in from the coasts and into the heart of cities. I picture it like one of those flipbook animations, the lines altered almost imperceptibly as the pages flutter down, and all the while the ragged outline of the raven inching closer.

I want to complete my raven map in the streets where I grew up. To find out how close the birds are to returning to central London and discover where the points I had plotted: the wild ravens flying in from the northern fringes of the Lee Valley; west from the Chilterns and south from Hampshire up into Croydon, intersect.

In his book, *Birds in London*, the Victorian ornithologist William Henry Hudson gives an authoritative account of where the last ravens nested in the capital. In the seventeenth century, he says, they remained a common sight, existing well after the red kite had been banished. They were easily identified from rural ravens by their dulled or dusty plumage, the result of foraging among the dust and ash heaps piled about the city. In modern London, you can still detect, in the silvery wings of some crows, the contemporary version of this poor urban diet of fried chicken and kebab scraps pilfered from the floor.

According to Hudson, ravens bred in Hyde Park until 1826, when a keeper pulled down the last nest with the young birds still inside. The adult pair fled, but the keeper kept one of the juveniles for himself. This tamed raven became well-known in the park, and would often hang about the workmen who were building John Rennie's stone bridge over the Serpentine. One day, a well-dressed woman was walking near the bridge when she felt a sudden peck on her ankle and looked down to see a raven tugging at her hems. In her panic, she dropped a gold bracelet, which the bird snatched in its beak and stowed away in a hollow tree in Kensington Gardens. The missing bracelet was searched for by the park authorities but never found.

Such was the raven's celebrity status that it was eventually stolen from the park. The bird returned a few weeks later, but 'moped a great deal' with its wings clipped. Sometime later it was found dead, floating in the Serpentine. According to Hudson, people thought he had drowned himself from grief at having been deprived of the power of flight.

In 1845, Hudson traced the last known breeding pair to what was then the northernmost outskirts of the city: Tottenham. The pair were nesting on one of the giant elms, which made up the Seven Sisters Road between Holloway and Tottenham, but a replanting programme led to all the old trees being cut down in 1852, and the raven's eyrie was gone for good. I grew up not far from the Holloway end and lived in a flat on Seven Sisters Road for several years in my late 20s. Nowadays, the trees are made up of a circle of seven hornbeams planted around Page Green near Tottenham in 1966, and three lanes of traffic thunders by.

Hudson's account of the last of the London ravens is verified by another naturalist of his era, the schoolmaster

Reginald Smith. In his *Bird Life and Bird Lore* he writes that, to the late-Victorian and Edwardian ornithologists, the raven seems to represent only loss: 'a state of things ... slowly passing away'.

As London's wild ravens grew diminished, so the fashion for keeping them as domesticated pets grew. According to Smith, young birds were sold in Leadenhall Market for 10 or 15 shillings apiece. The high price meant egg collectors would pillage every nest they could, and through doing so, further restrict the raven's range. The trend for keeping ravens publicly, championed by Dickens, chimed neatly with the Gothic Revivalism of the era. Some breweries kept ravens as mascots, so too did owners of stately homes – attracting the ravens back to their castles to buff up their own past and sense of self.

The most famous example of the birds being entwined in human history, as we all in this country know, is the story that if the resident ravens ever leave the Tower of London then the kingdom will fall and the ancient seat of power will crumble to dust. Popular myth, encouraged by the Tower, has always insisted that the ravens were first installed there by King Charles II and have remained in situ ever since. But in recent years this has been disproved, both by Geoffrey Parnell, formerly the official Tower of London historian, and the American author Boria Sax in his book, *City of Ravens*. Both have reached the same conclusions, that the ravens were introduced in the late-Victorian era to add gothic grandeur to Tower Green, the site where 10 noble prisoners, including three queens of England, were executed during the bloody century of Tudor rule. The ravenstone was, after all, an old English name for a place of execution.

Rather than the courtiers of Charles II, it is believed the ravens were provided by a firm called Philip Castang – an importer of exotic animals and foxes for hunting – whose owner wrote a letter to *Country Life* magazine in 1955 confessing he had supplied the first birds to the Tower. Despite the unequivocal evidence amassed by Sax and Dr Parnell, this has made little imprint on the public perception of the history of the Tower. The legend still persists in the guidebooks and is regularly trotted out in newspapers. This is the raven story we want to believe, and it seems nothing will change that – even if for a spell during the Second World War, the Tower was left with no ravens at all when the resident birds supposedly died of shock. When the Tower reopened to the public on 1 January 1946, ravens were miraculously back in situ, kronking out over the rubble of the city and a kingdom that had remained resolute.

* * *

I hear and read snatched reports of sightings in London from friends and birders: a raven seen over Tottenham Marshes; another possibly heard in Hampstead Heath; one confirmed near Paddington station and another at Christ Church in Spitalfields. This particular viewing was from an artist called Ian Harper, who is piecing together modern-day sightings of the same species that appeared in Thomas Bewick's 1832, *History of British Birds*. 'Perhaps they escaped from the Tower of London,' he wrote on his blog compiling the updated Bewick's compendium. 'It has been known.'

To discover if wild ravens are now back and actually breeding in London rather than making occasional forays through its congestion charge zones, I make an appointment to see the current Ravenmaster of the Tower, a man called Chris Skaife. Our modern interest in ravens seems to be confirmed by the huge social media following he has garnered, posting video clips and photographs of his seven birds online. Buzzfeed describes his job as 'insanely cool'.

We meet on a drizzly, grey London morning with gaggles of tourists in plastic rain ponchos clustering outside the Tower gates. Chris greets me at the entrance, an impressive figure in the navy blue and scarlet uniform of the Yeoman Warders, and leads me through Tower Green and into the grace-and-favour apartment inside the old castle walls he is given as part of his job. The living room is decorated with all manner of raven curios, collected by the 51-year-old, and presented by visitors who come from all over the world to see his birds.

In order to become a Yeoman Warder, one must first spend 22 years in the military and finish at the rank of warrant officer or above. Chris served in the Queen's Regiment (later the Princess of Wales's Royal Regiment) and rose to the rank of staff sergeant during a career that took him to Northern Ireland, Cyprus, Belize and Kosovo. He grew up in Dover and says he always had a keen love of history. After leaving the Army in 2005, he had no idea what else to do, so decided to try his luck at the Tower.

The Ravenmaster at the time was a man called Derek Coyle. By his own admission, Chris knew nothing really about the birds but took an interest in the castle residents. One evening, the Ravenmaster invited him down to the aviaries where the ravens sleep. 'He told me he thought the ravens might like me, so put me in a cage with two rather large male ravens in there,' Chris says. 'I thought they were looking at me to eat me but tried not to flinch and after a while, he could see I wasn't scared of them, so let me out the cage. After that, he took me on to his team and showed me how to look after them. I still always ask myself why he picked me.'

Chris became Ravenmaster in 2011. He talks me through his average day, watching the birds in their cages for sign of illness or injury ('ravens can fight to the death'), before letting them out to roam the grounds. The birds eat 170g (6oz) of raw meat daily, plus bird biscuits soaked in blood. They are fed chicks, mice, rabbit and berries. 'If they want a sweet during the daytime they just go and steal something off a child,' Chris says.

The legendary royal decree has it that six birds need to be resident in the Tower, although Chris keeps seven with one in reserve: Munin who is 22 (the oldest Tower raven lived to 44), Erin, Rocky, Grip, Harris, Jubilee and Merlina. When one of the birds dies – like in 2013 when an urban fox stole in and ate two – they are swiftly replaced. Sometimes, depending on the current state of the nation, and health of the Royal Family, the news of a bird's death is not immediately released for fear of exciting rumours that the kingdom could fall.

While Chris gets most of his birds from an unnamed breeder, the Tower started its own breeding programme in 1989 and that same year the first chick was hatched inside its walls. A competition was held among schoolchildren to name the bird, and it was decided on Ronald, short for Ronald Raven, after the US president at the time. The first ever British Prime Minister, Sir Robert Walpole, who moved into 10 Downing Street in 1735, was similarly nicknamed 'Robert the Raven', supposedly for his greed and ability to elicit money for tax purposes. One anarchic poem, *The Life and History of Robert the Raven*, printed in London in 1742 with the aim of destabilising the government, contains the lines:

Of all our product, trade and toil,
this raven wears the golden spoil;
he's black and fat, immensely rich,
and still the Eagle doth bewitch,
tho' he be rav'nous, most unclean,
he lurks and rules behind the screen.

By his own admission, Chris is no natural scientist, but listening to some of his raven stories; their personalities and intelligence; how they establish hierarchies and work together to attack other animals and trick humans; how pairs connect and how weaker birds are ruthlessly expunged, brings together so much of what I have heard and seen for myself. Merlina appears in most of the videos posted online because, by the time he arrived, she had already become 'humanised'. In 2011 the retired Ravenmaster visited the Tower in civilian clothing, and despite not having seen Merlina for seven years, Chris says she instantly recognised her former keeper and flew towards him across Tower Green.

Chris aims to keep his other birds as wild as possible. Some of them he has no choice. Munin, for example, loathes him because of a perceived slight he once committed as a junior assistant. 'If you get the trust of a raven, it will trust you for life,' he says. 'If you don't they will hate you for life.'

Traditionally, Tower ravens would have their wings clipped to stop them flying away, although a few still managed to escape. One raven named Grog who made his escape in 1981, was last spotted outside an East End pub called the Rose and Punchbowl. Nowadays, Chris only trims individual flight feathers by a third or a half, and this means that, even if they lack the agility of a truly wild raven, the birds could still fly away. In two years of trials, no bird has yet escaped for good. Merlina will fly across the Thames to a wharf on the other side of the river, but so far, she has always returned.

He regularly receives reports of potential raven sightings in London, although often it is a blurry image that could just as easily be a large crow. 'I've long said there is a distinct possibility there will be ravens flying around the heart of London in 10 years' time,' he says. 'And actually, that may end up happening even earlier.'

★ ★ ★

'Out into the outskirt's edges, where a few surviving hedges, keep alive our lost Elysium.' My train rattles away from the platform at London Bridge, and after passing the shiny spires of Canary Wharf, I am out into the suburban Metroland of John Betjeman's imagination. It is a Saturday morning and early enough for my carriage to be empty, aside from a few late-night stragglers who missed the last train home, and two women in their 20s who have just finished their long shift at a 24-hour casino in the West End. They gossip about work, tiredness blurring the edges of their conversation. As we head out east, the silences grow longer.

I am on my way to meet a birdwatching friend and avowed early riser, David Darrell-Lambert. The first time we crossed paths was at 4 am in Queen's Woods in Highgate, where I had gone to write an article about the urban dawn chorus. In his Tottenham accent, David pointed out goldcrests, tawny owls and blackbirds that had adapted to mimic wailing police sirens haring down the nearby A1. It was only after several hours in each other's company that the light grew bright enough to see our faces. David's had a broad, eager grin – ever present in all the times we have met over the years.

A talented recorder of birdsong, David has been conducting urban surveys around London since 1988, and in 1991 began illustrating the London Bird Report. He counts birds for everybody from the Queen at Buckingham Palace to local community interest groups. If anyone knows about the closest pair of ravens nesting to the centre of London, it would be him.

I step off the train at Greenhithe (for Bluewater) Station, next to the huge shopping centre on the southern bank of the Thames. David is waiting there with his sunglasses on. We head off in a car embossed with his company logo 'Bird Brain UK', listening to nineties hip-hop, en route for Swanscombe Marshes.

It is high spring now, but we made this same trip a few
months before with little success. Then, it was such a cold
morning that puddles on the ground had frozen solid and
the wind, singing over the mudflats, brought tears to my
eyes. The marshes are situated between Queen Elizabeth II
Bridge to the west and Tilbury Docks to the east. They are
currently being eyed up by developers who want to build a
theme park smack bang in the middle of what remains one
of the last few examples of a rare and precious wilderness so

close to the city, and one of the last sizeable areas of marshland in the lower Thames corridor. That bitter morning, as we unsuccessfully hunted for signs of ravens building a nest, we were pursued by an intimidating security guard in a 4x4 doing laps of a place that is, still for now, common ground. Sometimes, David tells me, they are accompanied by Alsatians.

A few months on, the future of Swanscombe remains in the balance with talk of construction crews soon to roll in, but at least this time the sun is shining, and David has heard that the ravens have not just built a nest but also reared three young. We strike out past the reedbeds stalked by kestrels and marsh harriers, catching occasional glimpses of bearded tits making a mad dash for cover. We pass a heronry some half a dozen nests-strong at the base of a patch of sycamores. Whitethroats scratch out their songs as they dart from wire to bush.

We follow the path down a line of pylons, the tallest of which is 204m (670ft) high, engineered to carry cable high over the Thames. All around is evidence of industry that has profited from this land: old chalk pits, cement works and warehouse buildings now home to thousands upon thousands of roosting gulls. By the path, empty beer cans have been skewered on to the branches of a juvenile alder, a glittering Kronenberg Christmas tree.

We search the pylon where the ravens are thought to have built their nest, but despite seeing the mass of sticks high up in the gantry, there is no sign of the actual birds. A familiar, unspoken disappointment creeps in between us, but then, from another pylon to our right, we hear the ravens call. We spot one bird in the air first and then follow it down to where its three young are waiting halfway up the steel frame, cawing loudly for food. The gape lines on either side of their mouths are still pale and the insides pink, reflecting the fact these birds are only a few months old. Their feathers are ruffled easily by the wind, fluffing them up as if they have been rubbed dry by a towel. Their legs are pale and have not yet developed the thick bloomers that distinguish a raven's nether regions. They call out in a higher, quicker tempo to the adult bird, but it is still unmistakably the sound of a raven. We take it in turns to look through David's telescope, breathless and silent. The young birds are learning to fly.

Over the next hour or so, we watch as the fledgling ravens leap off from their pylon perch and into the air. Compared to their parents, who keep an easy distance soaring above them, the young birds flap their wings far more eagerly to compensate for their undeveloped breast muscles. Each grateful landing back on the gantry is marked by a chorus of calls from their siblings. By and large, the adult pair keeps silent. In a few months, these young ravens will be gone from the nest, perhaps heading westward down the Thames towards the city?

I watch these young ravens buffeted by the wind like bees. Their flight lines falling and rising over the marsh as they build an understanding of the contours of their new world, the synapses in their powerful brains flickering into life as they develop maps of their own. All this must seem normal to them: the factory roofs, cranes, new-build apartments and distant roar of the M25, the only version of life they know. Intelligent as they are, and however wary their deep evolutionary impulses may make them, these young ravens cannot understand that these human developments do not simply stop where they are, but spread.

I picture all the life on this marsh and think back to the empty spaces on the Victorian maps of my own area and how quickly these things can disappear. It strikes me once more how the ravens need us, to continue to thrive; relying upon our tolerance and acceptance and desire to share spaces with them.

Watching the birds dive under the fizzing pylon wires, I also realise just how much we need them close by. To provide us with a glimpse of wildness in a world hell-bent on civilising its furthest reaches, while at the same time inching closer towards the abyss. The raven will always continue to represent our own projections. This modern omen remains as yet ill-defined; our shared futures unresolved.

Further Reading

Aldhouse-Green, Miranda, 1992. *Animals in Celtic Life and Myth*. Routledge, London.

Alexander, Michael, 1973. *Beowulf: a verse translation*. Penguin, London.

Anon, 1877. *The Raven's Feather*. T. Nelson & Sons, Edinburgh.

Betjeman, John, 1959. *Collected Poems*. Houghton Mifflin, Boston.

Blackburn, Julia, 1991. *Charles Waterton*. Century, London.

Brown, Ian, 2008. *Beacons in the Landscape: the hillforts of England and Wales*. Windgather, Oxford.

Burns, Robert, 1905. *The Cotter's Saturday Night*. Chatto & Windus, London.

Chapman, Abel, 1907. *Birdlife of the Borders on moorland and sea*. Gurney and Jackson, London.

Cocker, Mark, 2008. *Crow Country*. Ulverscroft, London.

Cramp, Stanley, 1980. *Handbook of the Birds of Europe, the Middle East and North Africa*. Oxford University Press, Oxford.

Davies, Walford and Ralph, Maud, 2003. *Dylan Thomas – Collected Poems*. Phoenix, London.

De Capel Wise, John Richard, 1895. *The New Forest: its history and scenery*. Gibbings, London.

Dickens, Charles, 1998. *Barnaby Rudge*. Wordsworth, London.

Dyer, John, 1930. *Grongar Hill*. Swan Press, London.

Emery, Nathan, 2016. *Bird Brain: an explanation of avian intelligence*. Ivy Press, Princeton.

Garner, Alan, 1967. *The Owl Service*. Collins, Great Britain.

Goodwin, Derek, 1976. *Crows of the World*. British Museum Natural History, London.

Heinrich, Bernd, 1999. *Mind of a Raven: an investigation into the mind of a raven*. Cliff Street Books, New York.

Heinrich, Bernd, 1991. *Ravens in Winter*. Vintage, London.

Hudson, William Henry, 1928. *Birds in London*. Duckworth, London.

Kelsall and Munn, 1905. *The Birds of Hampshire and the Isle of Wight*. Witherby and Co, London.

Lorenz, Konrad, 1970. *Studies in animal and human behavior*. Methuen, London.

Lovegrove, Roger, 2007. *Silent Fields: the long decline of a nation's wildlife*. Oxford University Press, Oxford.

Mabey, Richard, 2010. *The Unofficial Countryside*. Little Toller, Dorset.

Mackay Brown, George, 2014. *Beside The Ocean of Time*. John Murray, London.

Mackay Brown, George, 1989. *The Wreck of the Archangel*. John Murray, London.

Marzluff, John and Angell, Tony, 2005. *In the Company of Crows and Ravens*. Yale University Press, New Haven.

Monbiot, George, 2013. *Feral: rewilding the land, the sea, and human life*. Penguin, London.

Oliver, George, 1866. *Ye Byrde of Gryme*. A. Gait, Grimsby.

Palsson, Herman and Edwards, Paul, 1981. *Orkneyinga Saga: the history of the Earls of Orkney*. Penguin, London.

Poe, Edgar Allen, 1994. *Tales of Mystery and Terror*. Puffin, London.

Ratcliffe, Derek, 1997. *The Raven: a natural history in Britain and Ireland*. T. & A. D. Poyser, London.

Sax, Boria, 2011. *City of Ravens: the extraordinary history of London, the Tower, and its famous ravens*. Duckworth Overlook, London.

Scott, Walter, 1810. *Minstrelsy of the Scottish Border*. Anon.

Smith, Reginald Bosworth. 1905. *Bird Life and Lore*. John Murray, London.

Walsh, J. H., 1957. *The Poetry of John Clare. An Anthology*. Chatto & Windus, London.

Waterton, Charles, 1839. *Essays on Natural History*. Longman & Co, London.

Watson, Sally, 1991. *Secret Underground Bristol*. Bristol Junior Chamber, Bristol.

White, Gilbert, 1829. *The Natural History of Selborne*. Constable & Co., Edinburgh.

Woolfson, Esther, 2008. *Corvus: a life with birds*. Granta, Great Britain.

Acknowledgements

This book is the product of the generosity of others, who have given so much of their time and expertise assisting with my endeavours.

I would like to thank Joel and Bob Gilbert, Paul Stancliffe at the British Trust for Ornithology, Florence Wilkinson at Warblr, Keith Betton at the Hampshire Ornithological Society, Rob Shepherd, Andy Page, Paul Holt, Francis Hickenbottom and Lesley Taylor at the Wakefield Naturalists' Society, the staff at the West Yorkshire Archive Service and British Library, Bristol Ornithological Club, Michael Gibson, Mark Atherton, Historic Royal Palaces, Doug Simpson, David Darrell-Lambert, Chris Booth, Fran Flett Hollinrake and Andrew Hollinrake, Tom Muir, Sarah-Jane and Elliot Manarin, Nigel Brown, Igraine Skelton, Ian Rutherford, Roger Lovegrove, and, posthumously, Derek Ratcliffe, whose passion and knowledge of ravens is as inspiring as it is unparalleled and whom I would have loved to have met.

I am most grateful to my Bloomsbury editors Alice Ward and Jim Martin for their advice and unwavering enthusiasm for my ideas.

I am also indebted to my *Telegraph* editors Vicki Harper, Jessamy Calkin and Jane Bruton for all their support, and Liz Hunt for the years she spent beating my copy into shape.

My apologies to those other people I have no doubt missed out.

Finally to my family, friends and journalist colleagues whom I have bored and neglected in equal measure over the course of writing this book, my mum for her razor-sharp proofreading and Jon Day for his invaluable advice with final drafts.

All remaining mistakes are my own doing.

Index